Fife and Angus
Geology

Fife and Angus Geology

An Excursion Guide
Third Edition

A. R. MacGregor

The Pentland Press
Edinburgh – Cambridge – Durham – USA

First published in 1996 by
The Pentland Press Ltd
1 Hutton Close
South Church
Bishop Auckland
Durham

ISBN 1–85821–353–3

Typeset by Carnegie Publishing, 18 Maynard St, Preston
Printed and bound by Bookcraft (Bath) Ltd

Contents

DESCRIPTIVE ITINERARIES

Contents

Illustrations

Maps

Tables

Preface

Fife and Angus Geology was first published in 1968, a second edition appearing in 1973. These early editions owed much to the encouragement of the late Professor C. F. Davidson and the critical appraisal of R. Johnston, both of the St Andrews University Geology Department.

By 1985, however, Fife and Angus Geology was out of print and much new work on the geology of the area had been published. The opportunity was therefore taken to re-write completely the introductory chapters, incorporating new information into all the itineraries, omitting from the guide book the excursions to the area between Stonehaven and Aberdeen and to South Queensferry since both are very well covered by the Aberdeen and Lothian guides respectively, and combining or modifying some of the previous itineraries. Two new itineraries have resulted, one for the North Fife Hills and the other for the area lying between Dundee and Perth.

Over the years gentle but persistent encouragement to produce a new edition came from Douglas Grant and many other friends and colleagues and I am grateful to them all for this, as I am to my wife for reading and rereading the manuscript over a considerable period. The manuscript was put on the word processor by Sue Canfield and Margaret Connolly with patience and understanding. The maps and tables were entirely redrafted or are new, the work of Graeme Sandeman and Janet Mykura of the Cartographic and Graphic Services of the School of Geography and Geology, University of St Andrews.

Acknowledgements

I am grateful for the financial support of North East Fife District Council, the Edinburgh Geological Society, the Geological Society of Glasgow, and the Geology Department of the University of St Andrews.

Itineraries in the guidebook are based partly on the literature referred to after each excursion, partly on the author's own field work and partly on information freely supplied over many years by his colleagues, past and present, in the Department of Geography and Geology, St Andrews University. The chapter on the Quaternary has benefited greatly from the advice of Professor C. Ballantyne. Earlier editions benefited from discussions with Professor E. H. Francis and Messrs I. H. Forsyth and J. I. Chisholm of the Geological Survey when they were re-mapping much of East Fife for the North Berwick and Arbroath map sheets for the Geological Survey. For this revised edition, David Walker walked almost all the excursion itineraries and I am grateful for his helpful comments.

The following maps are based, at least in part, on previously published work: maps 2, 3, 4, 7, 12, 21, 22 and 23 on maps of the Geological Survey; map 5 on the work of D. Peacock; map 6 on the work of B. Harte; maps 8 and 9 on the manuscript maps of the late W. T. Harry; map 11 on the work of C. Rice; map 14 on a map of D. Balsillie; map 15 is after S. R. Kirk and maps 18 and 19 after Francis and Hopgood.

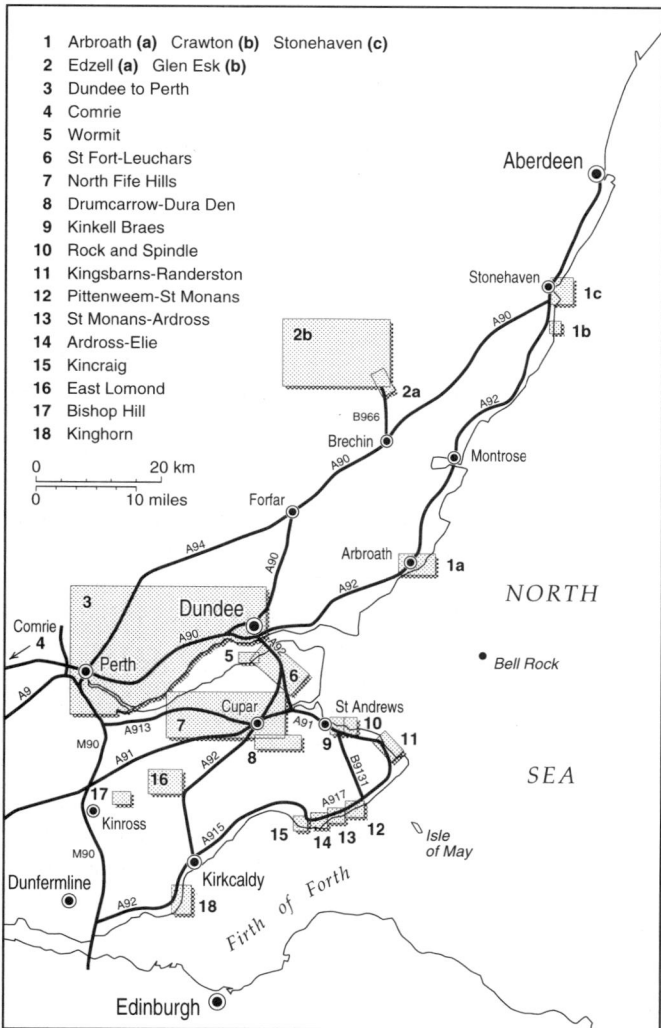

1   Arbroath **(a)**   Crawton **(b)**   Stonehaven **(c)**
2   Edzell **(a)**   Glen Esk **(b)**
3   Dundee to Perth
4   Comrie
5   Wormit
6   St Fort-Leuchars
7   North Fife Hills
8   Drumcarrow-Dura Den
9   Kinkell Braes
10   Rock and Spindle
11   Kingsbarns-Randerston
12   Pittenweem-St Monans
13   St Monans-Ardross
14   Ardross-Elie
15   Kincraig
16   East Lomond
17   Bishop Hill
18   Kinghorn

MAP 1: Excursion location map.

2

Introduction

Geographical Setting

The area covered by the guide book is bounded by a line extending north-west from Kinghorn to Perth and then north-east to the coast at a point just north of Stonehaven, then back down the coast to St Andrews, Fife Ness and along the south coast of Fife to Kinghorn. Two excursions lying outside this area, those to Glen Esk and to Comrie, are included to broaden the range of geology readily accessible from St Andrews.

The area includes the south-east corner of the Highlands where a number of glens open onto Strathmore, e.g. Glen Esk and Glen Clova. Strathmore is bounded on its south-east margin by the Sidlaw Hills extending north-eastwards from Perth to Forfar, and those in turn are separated from the geologically similar Ochil Hills to the south-west by the Tay Valley at Perth and by Strathearn slightly further south. The North Fife Hills form a continuation of the Ochil Hills and are separated from the rest of Fife by Stratheden, extending north-east from Loch Leven to the sea near Leuchars. In the southern part of East Fife the hills are relatively isolated, e.g. the Lomond Hills, east of Loch Leven, and Kellie Law and Largo Law further east, while the Binn of Burntisland dominates the south-west corner of the area.

Geological Setting

The geological control of the topographic pattern soon becomes apparent when the geology and topography are considered together. The Highlands bounding Strathmore to the

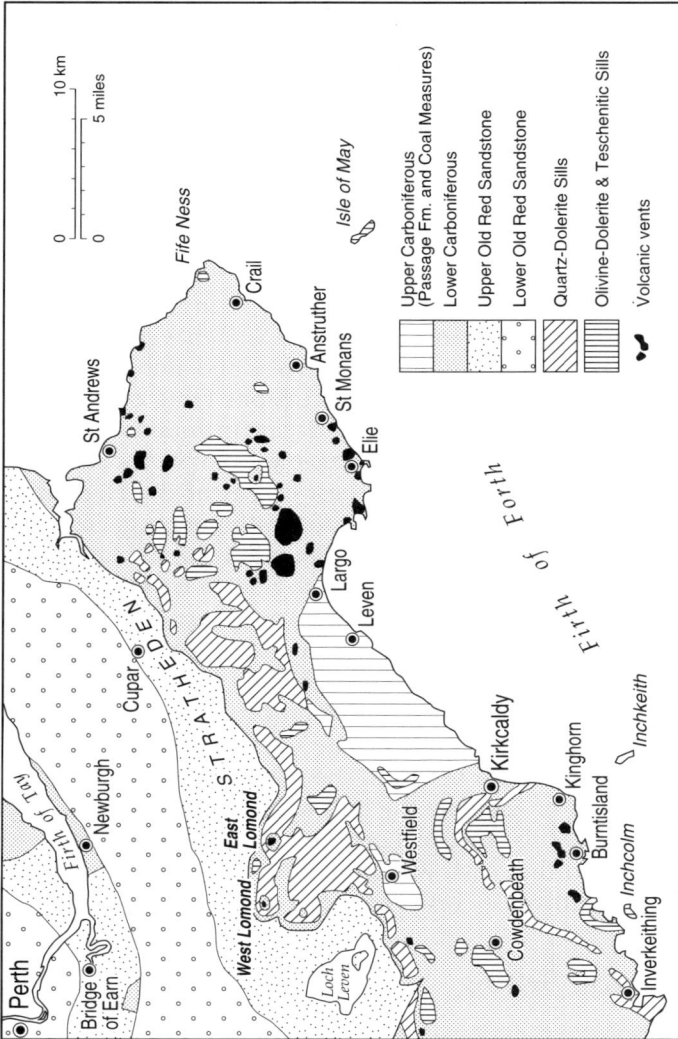

MAP 2: Simplified geological map of Fife.

Upper Carboniferous
(Passage Fm. and Coal Measures)

Lower Carboniferous

Upper Old Red Sandstone

Lower Old Red Sandstone

Quartz-Dolerite Sills

Olivine-Dolerite & Teschenitic Sills

Volcanic vents

north-west are composed of durable, old metamorphic rocks and are separated from the rest of the area by the Highland Boundary Fault, a great fracture running south-westwards right across Scotland from Stonehaven to the Clyde at Helensburgh. Strathmore is occupied by Old Red Sandstone rocks in a large syncline running parallel to the fault, the Strathmore Syncline. The Sidlaw Hills and the North Fife Hills are composed primarily of lavas of Lower Old Red Sandstone age on either side of the Sidlaw Anticline – another NE–SW trending structure. The resistant lavas have withstood weathering and erosion to form two ranges of hills that can be traced south-westwards until they meet in the Ochil Hills.

The low ground of the Carse of Gowrie, between the

FIGURE 1: The Lomond Hills seen from the Bow of Fife. The peaks are the eroded remnants of two Carboniferous-Permian volcanic necks now standing above the general level of the Midland Valley Sill which forms the marked scarp beneath the summit of the West Lomond on the right. The Howe of Fife in the foreground is covered by fluvioglacial deposits overlying horizontal Upper Old Red Sandstone sediments.

Era	System	Group etc.	Maximum thickness in metres	Igneous Rocks Eruptive	Igneous Rocks Intrusive	Age in My
Tertiary & Mesozoic / Quaternary	—	Glacial, fluvioglacial and marine	60+	—	—	2
	—	—	—	—	—	225
	Permian	—	—	—	necks	286
Upper Palaeozoic	Carboniferous	Coal Measures	1100+	minor intermittent volcanics	Quartz-dolerite sills & dykes	
		Passage Formation	270+		Olivine-dolerite, teschenite sills	
		Upper Limestone Formation	335+			
		Limestone Coal Formation	430		necks	
		Lower Limestone Formation	175			
		Strathclyde Group	2000	Burntisland volcanics		
		Inverclyde Group	180+	—		360
	Devonian	Upper Old Red Sandstone	700±	—	—	
		Lower Old Red Sandstone	7000	Ochil, Fife, Sidlaw, Montrose volcanics	Dundee sills N. Fife	410
Lower Palaeozoic	Silurian	Stonehaven Gp	1500	—	Newer Granites	
	Ordovician	Highland Border Complex	?	spilitic lavas	serpentinite	—
	Cambrian	—	—	—	—	590
Pre-Cambrian	—	Dalradian — Southern Highland Group	2000+	—	Epidiorites	

TABLE I: Stratigraphic succession for the Fife and Angus area.

Sidlaws and the North Fife Hills, is occupied mainly by Upper Old Red Sandstone sediments, let down between the North and South Tay Faults in the centre of the Sidlaw Anticline. These in turn are overlain by thin, very much younger, Lateglacial and Postglacial clays in the Tay Estuary.

Stratheden is also underlain by soft Upper Old Red Sandstone sediments and is bounded to the north by the more resistant Lower Old Red Sandstone lavas of the North Fife Hills. The southern slopes are composed of Carboniferous rocks, which are also relatively soft. These, however, are protected from erosion by the large Midland Valley Sill which forms such scarp features as Walton Hill near Cupar, the Lomond Hills and Bishop Hill. Standing above this scarp are the peaks of the East and West Lomonds, twin volcanic necks, now little more than deeply eroded stumps.

Much of the higher ground between St Andrews and Leven is also capped by dolerite sills, but Kellie Law, Largo Law and Kincraig at Elie are three other old volcanic necks (see Map 2).

The low ground between Largo and Kirkcaldy is underlain by Upper Carboniferous Coal Measures, soft rocks with few natural exposures, but giving rise to good farm land. Westwards high ground reappears behind Kinghorn and Burntisland where it is due largely to volcanic rocks, both lavas and necks of old volcanoes of Lower Carboniferous age. Benarty, south of Loch Leven, is capped by a continuation of the Midland Valley Sill bordering the south side of Stratheden. The prominent headland at North Queensferry at the north end of the Forth bridges is also part of the Midland Valley Sill, while the Isle of May at the entrance to the Firth of Forth is a tescherite sill.

Glacial deposits, both till and gravels and sands, are widespread and the latter locally give rise to characteristic land forms, e.g. at Loch Leven, Collessie, Leuchars and Forfar. The passage of ice has 'rounded the corners and smoothed the edges' over much of the area though this effect is gradually disappearing through post-glacial weathering and erosion.

This smoothing is in marked contrast to the deep glacial erosion in parts of the Highlands.

Scope of the Guide and further references

The whole-day and half-day excursions are designed to illustrate the geology of the area. Most of them are or were excursions run from the Department of Geography and Geology, St Andrews University. Two can be reached from St Andrews on foot; the remainder require transport and in a number of cases it is helpful to be dropped at one point and picked up at another.

References to the appropriate maps and literature are included with each itinerary, but for the area as a whole the 1/250,000 Tayforth Solid Geology Sheet 56N 04W of the British Geological Survey is useful. One-inch to the mile or 1/50,000 Geological Survey maps are available for most of the area and are listed at the beginning of each itinerary with the prefix GS. Six-inch to the mile or 1/10,000 geological maps of some of the coalfields are published; others are available for inspection at the office of the British Geological Survey in Edinburgh. Out of print maps may also be consulted there. Ordnance Survey maps are also listed for each itinerary with the prefix OS and refer to the 1/50,000 maps of Scotland.

The British Regional Geology Handbooks *The Midland Valley* by I. B. Cameron and D. Stephenson (1985) and *The Grampian Highlands* by G. S. Johnstone (1966), both published by HMSO, contain abundant references and are revised from time to time. For a treatment of the whole country the *Geology of Scotland* edited by G. Y. Craig (1991) is comprehensive and has long lists of references. Readers not familiar with geological terminology are referred to the Penguin or other geological dictionaries. The names of fossils have been kept to an elementary level deliberately in the belief that names such as *Productus* are more meaningful to the non-specialist. Full faunal lists for most of the fossil localities in Fife are to be found in the publications of the Geological Survey, e.g. *Geology of East Fife* (Forsyth and Chisholm 1977).

Map 1 indicates the location of the 18 excursions. Coastal excursions almost all require mid- to low tide and for Excursion 15 (Kincraig) low tide is essential for access.

The Geological Code of Conduct

As the number of people interested in geology has increased over the years, so the number of visits to localities with well displayed geology has increased. It is thus essential that anyone looking at rocks in the field should be aware of the code of conduct This is published by the Geologists' Association, Burlington House, Piccadilly, London, W1V 0JU. From it the following points are here stressed:

- All inland exposures are on land belonging to a landowner whose livelihood and that of others can be affected by your behaviour. It is important and a courtesy that permission should be sought before crossing farmland. Do not walk through standing crops. Gates should be shut and stiles used wherever possible.

- In working quarries it is essential, and indeed required by law, that permission to enter is sought beforehand and a safety helmet worn.

- Common sense points to care and caution near steep slopes and cliffs. When collecting, if this is necessary, care must be taken to leave the site in a clean and tidy state, i.e. in the condition that you would wish to find it yourself. When hammering rocks remember those around you: many accidents have been caused by rock splinters. Already fallen rocks often provide good samples without having to dislodge more material. Sketches and photographs often provide much of the information wanted.

- Do not collect from walls or undermine buildings. Do not interfere with machinery, livestock or crops; they are all part of the livelihood of someone in the vicinity.

- On coastal excursions it is important to know the state of the tide, both to ensure exposure of the rocks and also for safety. This is particularly important on Excursion 15 to Kincraig, Elie. A tide table is included to allow timing of excursions.

- Sites of Special Scientific Interest, known as SSSIs, are designated by Scottish Natural Heritage and hammering of rocks is not permitted in these. Loose material on the beach may, however, be collected. Much of the Fife coast is covered by SSSIs, in particular the following stretches including the itineraries within them: Burntisland to Kirkcaldy; East Wemyss to Anstruther; St Andrews to Boarhills; and Wormit to Balmerino. Inland in Fife, Bishop Hill is also a SSSI. North of the Tay, Whiting Ness at Arbroath is a SSSI as are Den of Fowlis and the Gallowflat Claypit at Pitfour.

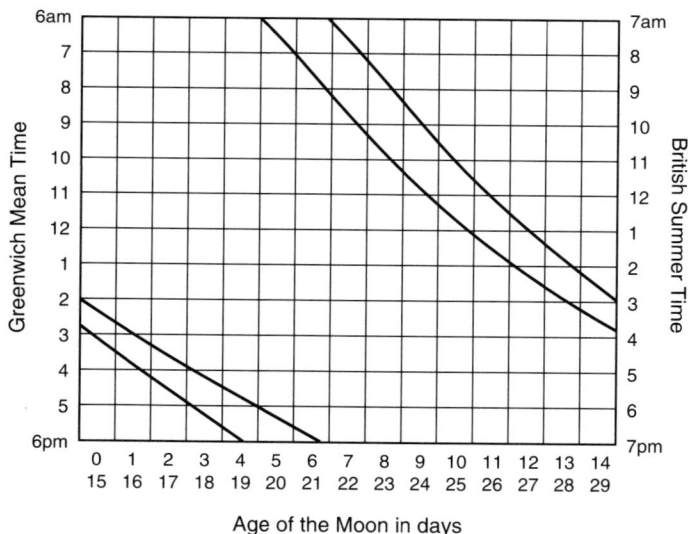

Age of the Moon in days

DUNDEE TIDE TABLE.

10

Tides

To determine the time of high tide on any particular day, note how many days after new moon or full moon it is. New moon is marked on the chart as day zero and full moon as day 15. Then read along the bottom of the chart to the day in question and follow the column up. The time of high tide lies between the two heavy black lines which cross the chart obliquely. Spring tides, those with the greatest range, occur approximately 2 days and 16 days after new moon; while neap tides, those with the least range, occur approximately 9 and 23 days after new moon.

High tide is 60 minutes earlier at Stonehaven.

High tide is 50 minutes earlier at Fife Ness.

High tide is 40 minutes earlier at Methil.

More precise times can be obtained from the Admiralty Tide Tables.

Geology of the area

The general geological succession in the area covered by the guide book is given in Table I.

Chapter 1

Dalradian

Table II
Excursions 2 and 4

Within the area covered by the guide book the oldest rocks present belong to the Dalradian Supergroup, a suite of metamophosed sedimentary and igneous rocks, now known to be entirely Precambrian in age (Rogers *et al.* 1989). Much general background information on the Dalradian can be found in Harris and Pitcher (1975) and in Johnson (1991), but the view expressed in both these publications that the Dalradian extends from the Precambrian up into the Cambrian is no longer tenable.

Throughout most of the area the Dalradian sediments consist principally of shales or slates and quartzo-feldspathic grits in which graded bedding is the commonest sedimentary structure. They belong to the Southern Highland Group, the uppermost group of the Dalradian, and have generally been correlated with the Aberfoyle Slates and Ben Ledi Grits of the Perthshire succession. However, in Upper Glen Esk older parts of the Dalradian are developed and extend down to the underlying Tarfside Limestone, assigned to the Argyll Group of the Dalradian (Harte 1979).

Structure

The Dalradian rocks have a broad regional NE–SW strike and have been subjected over a prolonged period of time to more than one phase of folding (see Table II).

Expressing this in very simple terms: a major, early, recumbent fold, the Tay Nappe of D1/D2 age, was overturned to the south-east until the limbs were approximately horizontal. Much later (D4), this fold, and other similar folds in the South-East Highlands, were refolded so that the front of the Tay Nappe was bent down to the south-east.

This has led to two contrasting structural areas on the ground. Further to the north-west, and lying some kilometres north-west of the Highland Boundary Fault, is the 'flat belt' in which grits and slates are disposed in folds with long, nearly flat, limbs and short, steep limbs. Adjacent to the Highland Boundary Fault is the 'steep belt', an apparently monotonous sequence of steeply dipping grits and slates. The boundary between these two belts is known as the 'downbend' and marks the axial plane of a major late D4 fold (Harte *et al.* 1984, figs 3, 4, p. 156).

By using graded bedding to determine the correct way up it has been possible to demonstrate that in the steep belt the rocks are isoclinally folded and, in addition, overturned to the south-east.

On the Stonehaven coast the downbend, the change from the flat belt to the steep belt, takes place over a few hundred metres some 3km north of the Highland Boundary Fault. South-westwards the downbend gradually diverges from the Highland Boundary Fault until it lies 5–10km to the north-west, e.g. in the Comrie area.

Harte *et al.* (1984) have summarized in more detail much research work particularly for Glen Esk (Excursion 2), but also for Comrie (Excursion 4) in a geological succession involving an earliest D1/D2 set of structures of which the Tay Nappe and the underlying Tarfside Nappe are the principal features in this area. The Tay Nappe is a major recumbent anticline closing to the south-east in which the inverted limb is coincident with the surface of the ground except in Upper Glen Esk where the Tarfside Culmination, a later D3 or even D4 structure, has brought up to higher levels the 'right-way-up' rocks of the underlying Tarfside Nappe.

Metamorphism

Not only are the rocks highly folded, they have also undergone major regional metamorphism of varying intensity. It is believed that this developed gradually after the D2 deformation and reached its peak during the D3 period of deformation. This metamorphism was first described by Barrow (1893, 1912).

The low grade chlorite and biotite metamorphic zones adjacent to the Highland Boundary Fault are rapidly succeeded to the north-west by the garnet, staurolite and kyanite zones and the highest, the sillimanite, zone is recorded on the east coast within 7km of the fault. This is in marked contrast to the wide metamorphic zones 80 to 150km to the south-west. Two contributory causes for the narrowness of the metamorphic zones have been recognised. Firstly, the zones, originally near to flat lying, have been rotated through nearly 90° at the D4 downbend, which is younger than the metamorphism, so that they are now approximately vertical. Secondly, even allowing for the D4 downbend, Harte *et al.* (1984, pp157–8) have drawn attention to the very high thermal gradients pointing to a rapid drop in temperature to the south-east at the time of metamorphism. This they attribute to possible downfaulting on the south-east at that time. The cause of metamorphism is still disputed, but Harte and Hudson (1979, p333) suggest that 'regional metamorphic gradients, migmatites and older granite bodies are all related to extensive regional magma intrusion in the deep crust'.

Isotopic age dating has allowed a timetable of events to be worked out, thus permitting correlation of periods of deformation, intrusion of igneous bodies and establishment of metamorphic zones. These are summarized in Table II.

During the late D4 movements retrogressive metamorphism took place, especially near the downbend and the axes of minor D4 folds, white mica, biotite and chlorite being produced (Harte *et al.* 1984, pp153–4). More detailed comments on the appearance of the rocks are contained in the itineraries for Excursions 2 and 4.

TABLE II. Dalradian Evolution
(After Rogers *et al.* 1989 and Harte *et al.* 1984)

Time (My)	Structures	Metamorphism	Uplift	Igneous Events
pre 590	D1/D2 Tay and Tarfside Nappe	Low grade		
550				
500	D3	Peak metamorphism	Local in Upper Glen Esk	
450	D4 Highland Border Downbend	Retrogressive Metamorphism	Regional Uplift	
400				Post-tectonic Granites, e.g. Comrie

References

BARROW, G., 1893. On an intrusion of Muscovite-biotite Gneiss in the South-Eastern Highlands of Scotland, and its accompanying Metamorphism. *Quart. Jour. Geol. Soc. Lond.*, **49**, 330–58.

————— 1912. On the Geology of Lower Deeside and the Southern Highland Border. *Proc. Geol. Ass. Lond.*, **23**, 274–90.

HARRIS, A. L. and PITCHER, W. S., 1975. The Dalradian Supergroup. In HARRIS, A.L. *et al.* A correlation of the Precambrian rocks of the British Isles. *Geol. Soc. Lond. Spec. Rep. 6.*

HARTE, B., 1979. The Tarfside succession and the structure and succession of the eastern Scottish Dalradian rocks. In HARRIS, A.L. *et al.* The Caledonides of the British Isles – reviewed 221–8. *Spec. Pub. Geol. Soc. Lond. 8.*

——————— and HUDSON, N.F.C., 1979. Pelite facies series and the temperatures and pressures on Dalradian metamorphism in E. Scotland. In Harris, A.L. *et al.* The Caledonides of the British Isles – reviewed 323–37. *Spec. Pub. Geol. Soc. Lond. 8.*

——————— *et al.,* 1984. Aspects of the post-depositional evolution of Dalradian and Highland Border Complex rocks in the Southern Highlands of Scotland. *Trans. Roy. Soc. Edinb., Earth Sci.,* **75,** 151–63.

JOHNSON, M.R.W., 1991. Dalradian. In CRAIG, G.Y. (ed.) *The Geology of Scotland,* Geological Society, London, 125–60.

ROGERS, G., DEMPSTER, T.J., BLACK, B.J. and TANNER, W.G., 1989. A high-precision U-Pb age for the Ben Vuirich Granite: implications for the evolution of the Scottish Dalradian Supergroup. *Jour. Geol. Soc. Lond.* **146,** 789–98.

Chapter 2

The Highland Border Complex

Excursions 1 and 2

In the Fife and Angus area, that part lying within the Midland Valley of Scotland is underlain by Upper Palaeozoic rocks belonging to the Devonian and Carboniferous Systems, part of the Midland Valley Block of Curry *et al.* (1984). Beyond the Midland Valley to the north-west lie the Scottish Highlands, underlain by much older Dalradian rocks of the Precambrian, the Dalradian Block of Curry *et al.* (1984). These two major areas are separated by the Highland Boundary Fault, not a simple single fault plane but rather a series of anastomosing faults between which occur fault-bounded wedges of rock seldom more than a kilometre wide. These wedges are largely composed of the rocks of the Highland Border Complex which were, until recently, thought to have an affinity with the adjacent Dalradian (e.g. Harris and Pitcher 1975).

More recent work indicates that the Highland Boundary Fault is a major boundary separating 'terranes' originally far from their present position and brought together by means of very large horizontal or transform displacements. Caught up within this major tectonic boundary are the wedges or 'slivers' (Curry *et al.* 1984) assigned to the Highland Border Complex.

Within the area covered by the guide book, fossils have been obtained from rocks of the Highland Border Complex at three levels: in Glen Esk (1) in the Margie Limestone – Upper Ordovician (Burton *et al.* 1984) and (2) in 'greenstones' – Lower Ordovician (Downie *et al.* 1971, p24), and at Stonehaven in cherty shales and interbedded greenstones/volcanics – probably Middle Ordovician (Curry *et al.* 1984).

Modern stratigraphical practice (Curry *et al.* 1984) suggests that Barrow's (1901) terms Jasper and Greenrock Series and Margie Series as applied to the North Esk and Stonehaven sections should be dropped and the informal terms 'rock assemblage' used until such time as more closely defined terms can be set up. The terms assemblages 1, 3 and 4 allow comparison with the better documented Aberfoyle area and are used below with the older Barrow terms in brackets for convenience. A wide variety of rock types are found in Rock Assemblage 1 including serpentinite; Rock Assemblage 3 contains black shales, cherts and volcanics, e.g. pillow lavas (Barrow's Jasper and Greenrock Series), and finally Rock Assemblage 4 comprises breccias, conglomerates and arenites and, in Glen Esk, limestone (Barrow's Margie Series).

Of the Highland Border Complex the following can be said (Curry *et al.* 1984):

1. The complex is younger than both the Tay Nappe and the peak of regional metamorphism in the Dalradian (see Table II).

2. The complex is made up of a wide range of lithologies brought together in a series of fault slivers, and ranging in age through most of the Ordovician and according to Rogers *et al.* (1989) Lower Cambrian.

3. Omitting for the moment the Cambrian, the Highland Border Complex has a varied stratigraphical history comprising an older pre-Arenig serpentinite complex, overlain unconformably by carbonates (in the Aberfoyle area), conglomerates, black shales, cherts and volcanics of Lower and Middle Ordovician age. These in turn are overlain unconformably by Upper Ordovician arenites, shales, limestones and conglomerates.

4. The sequence occurs along the important boundary between a poorly known Midland Valley basement, a possible source of much of the Highland Border Complex sediments, and the Dalradian rocks of the Highlands.

The rocks of the Highland Border Complex may well have been formed in an oceanic marginal basin with the sediment source on the south-east side. Of the adjacent Dalradian rocks there is little or no evidence until near the end of the Lower Palaeozoic, a time of Dalradian uplift and retrogressive metamorphism and the formation of the Highland Border downbend (see Table II), and it seems likely that only at this time did the rocks of the Dalradian and the Highland Border Complex come together as a result of major transcurrent faulting (Harte *et al.* 1984, p162). Such major transcurrent faulting may also have been responsible for the introduction of the Cambrian Leny Limestone to the Callander area further west.

References

BARROW, G., 1901. On the occurrence of Silurian (?) rocks in Forfarshire and Kincardineshire along the eastern boundary of the Highlands. *Quart. Jour. Geol. Soc. Lond.*, **57**, 328–45.

BURTON, C. J. *et al.*, 1984. Chitinozoa and the age of the Margie Limestone of the North Esk. *Proc. Geol. Soc. Glasg.*, **124/125**, 27–32.

CURRY, G. B. *et al.*, 1984. Age, evolution and tectonic history of the Highland Border Complex, Scotland. *Trans. Roy. Soc. Edinb., Earth Sci.*, **75**, 113–33.

DOWNIE, C. *et al.*, 1971. A palynological investigation of the Dalradian rocks of Scotland. *Rep. Inst. Geol. Sci. Lond.*, No. **71/9**.

HARRIS, A. L. and PITCHER, W. S., 1975. The Dalradian Supergroup. In HARRIS, A. L. *et al.*, A correlation of the Precambrian rocks of the British Isles. *Geol. Soc. Lond. Spec. Rep.* 6.

HARTE, B. *et al.*, 1984. Aspects of the post-depositional evolution of Dalradian and Highland Border Complex rocks in the Southern Highlands of Scotland. *Trans. Roy. Soc. Edinb., Earth Sci.* **75**, 151–63.

ROGERS, G *et al.*, 1989. A high-precision U-Pb age for the Ben Vuirich granite; implications for the evolution of the Scottish Dalradian Supergroup. *Jour. Geol. Soc. Lond.*, **146**, 789–98.

Chapter 3

Devonian
(Old Red Sandstone)

Table III
Excursions 1, 2, 3, 5, 7, 8 and 17

T he Devonian rocks of the area are immensely thick. Often red in colour – they were deposited 20°–30° S of the equator (Rayner 1981) – and entirely non-marine, they fall into the facies known in Britain as the Old Red Sandstone – and this term, though only broadly equivalent to Devonian in a time sense, is used throughout this account. Unfortunately fossils are rare so that detailed subdivision of the Old Red Sandstone is difficult, but enough fossil fish have been found to show that only rocks belonging to the Upper and Lower divisions occur in the area and that there is a marked unconformity between them. It is, however, now generally agreed that the lowest units of the Old Red Sandstone at Stonehaven are late Silurian in age (House *et al.* 1977, p72).

Since Lower Old Red Sandstone rocks occupy approximately two thirds of the surface area included in the guide they are considered in some detail. Within the area they lie within the NE–SW trending Strathmore Syncline and Sidlaw Anticline and are largely cut off to the north-west by the Highland Boundary Fault and to the south-east by an unconformable cover of Upper Old Red Sandstone or by the sea.

3a. Lower Old Red Sandstone
(Excursions 1, 2, 3, 5 and 7)

Only in the Stonehaven district is the base of the Lower Old Red Sandstone seen (Excursion 1, Location 13) but its occurrence there enables the thickness of these beds to be determined as about 9km including volcanic rocks, both lavas and pyroclastics (Armstrong and Paterson 1970). The extreme rarity of fossils in the sediments and consequent absence of close palaeontological control for correlation has until recently led to difficulties in correlation within the area and much more so in correlation with elsewhere (House *et al.* 1977). These have been well reviewed by Mykura (1991).

Fossil fish and eurypterids are known from the Stonehaven Group at Stonehaven and also occur higher up in the succession in the Arbuthnott Group (Armstrong and Paterson 1970, pp10–13). Plant fossils occur in the Arbuthnott and Strathmore Groups (Armstrong and Paterson 1970, pp12,16). These too suggest that it is likely that the great thickness of beds in this area is equivalent to the Downtonian to Dittonian of the Welsh Borders, i.e. infra-Gedinnian to Emsian of the Standard Succession (House *et al.* 1977, Table II).

(1) Kincardineshire (largely after Armstrong and Paterson 1970)

The Stonehaven Group (1550m), the basal Old Red Sandstone, lies unconformably on the Ordovician Highland Border Complex. In the Cowie Formation (730m) basal pebble beds and breccias (Excursion 1, Location 13), with one thin lava flow, are succeeded after 90m by several hundred metres of alternating brown sandstones and reddish mudstones with mud cracks and rare burrows (Excursion 1, Location 11). These are followed by 'volcanic' conglomerates (i.e. with a predominance of volcanic rocks among the pebbles) which are in turn succeeded by grey sandy shales, yielding *Dictyocaris*, and then by tuffs and tuffaceous sandstones. The fish fauna, which includes *Hemiteleaspis*, *Pterolepis* and *Traquairaspis*, collected near the top of this succession, indicates a Downtonian age,

Ages	Stages	Kincardine-shire	Perth	Dundee	Fife
360 My top of ORS	FAMMENIAN	Upper Old Red Sandstone (<60m)	↑	Clashbenny Fm (600 - 900m)	Stratheden Gp — Knox Pulpit Fm (170m) / Dura Den Fm (30m) / Glenvale Fm (500m+) / Burnside Fm (120m+)
375 My base of U.ORS	FRASNIAN				
	GIVETIAN	Missing	Missing	Missing	Missing
	EIFELIAN				
Base of middle ORS (385 → My)	EMSIAN	Strathmore Gp (2000m)	Teith Fm (1200m+) Cromlix Fm (300 - 450m)		
	SIEGENNIAN	Garvock Gp (1500 - 2000m)	Scone Fm (1900m)	Arbroath Ss Fm (360m+) Auchmithie Cgl (0-240m) Red Head Fm (450m)	(sandstones)
	GEDINNIAN	Arbuthnott Gp (2000 - 3000m) Crawton Gp (670m) Dunnottar Gp (1600m)	Ochil Volcanic Fm (2400 - 90m)	Dundee Fm (2000m) Montrose Volcanic Fm (40m)	Ochil Volcanic Fm (2000m)
400 → My	INFRA-GEDINNIAN = PRIDOLIAN (U. Silurian)	Stonehaven Gp (1550m)			
		Highland Border Complex (Ordovician)			

TABLE III: Correlation table for the Old Red Sandstone of Kincardineshire, Tayside and Fife.

i.e. late Silurian or infra-Gedinnian, for these beds (House *et al.* 1977, p72). The Carron Formation (820m) comprises red-brownish cross-bedded sandstones with abundant volcanic detritus, conglomerates and some agglomerate.

The Dunnottar Group (1600m) mainly comprises conglomerates with some tuffaceous sandstones, while near the top are some basic andesite and basalt flows. The conglomerates in this group consist predominantly of pebbles and boulders, up to as much as 1m across, of mainly metamorphic rocks especially quartzite. In the belief that these clasts were derived directly from the Dalradian rocks on the north-west side of the Highland Boundary Fault such conglomerates were named by Campbell (1913) 'highland' conglomerates and the term has been retained. Such conglomerates have been discussed in detail by Haughton (1988, 1990).

The succeeding Crawton Group (670m) contains a higher proportion of lavas, but both volcanic and highland conglomerates also occur in addition to sandstones (Excursion 1, Locations 2–5). Near the Highland Boundary Fault andesite is a common lava while near the coast basalt is much more abundant including the well known Crawton Basalts with their very large feldspar phenocrysts. Boulders as much as 1m in diameter are not uncommon in some of the conglomerates. Haughton (1988, 1990) has demonstrated that whereas the Dunnottar and Crawton conglomerates on the north-western side of the Strathmore Syncline have a northerly source, the Crawton conglomerates on the south-eastern side of the Strathmore Syncline have a now concealed source to the south and east within the Midland Valley. The Lintrathen Porphyry (Excursion 2, Location 4) is an ignimbrite occurring at about the same horizon as the Crawton Basalts and cropping out on both sides of the Highland Boundary Fault between Glenesk and Dunkeld.

The Arbuthnott Group (2100m) shows differences on either side of the Strathmore Syncline similar to those displayed by the underlying Crawton Group. On the south-east side a great thickness of highland conglomerates, the Johnshaven

Formation, passes south into hypersthene-bearing andesites and basalts, the Montrose Volcanic Formation, resting on a highland conglomerate. On the north-west side the group is largely cut out along the Highland Boundary Fault.

The succeeding Garvock Group (1525m), which has its type area in the Forfar-Brechin area in Angus, comprises highland conglomerates, sandstones and flows of basalt and andesite. The sandstones frequently carry calcareous grains believed to be derived from the erosion of carbonate-rich soils (calcretes). The Pittendreich Limestone of localised but widespread occurrence near the top of the Garvock Group is such a calcrete (Armstrong and Paterson, 1970 p14).

The youngest group, the Strathmore Group (1800m) is free of volcanic rocks and in it conglomerates are restricted to the north-west side of the Strathmore Syncline. Bright red marls and red cross-bedded sandstones predominate, the latter becoming coarser to the north-west. The group is well exposed in the North Esk section at Edzell (Excursion 2, Locations 1–3) and has yielded the plants *Psilophyton* and *Arthrostigma* in the upper part of the group.

When the Kincardineshire succession is traced to the south-west along the Highland Boundary Fault the Stonehaven and Dunnottar Groups are rapidly overlapped and the Crawton Group is much reduced so that the Lintrathen Porphyry at the top of the group rests unconformably on the Dalradian at Alyth and Dunkeld (Armstrong and Paterson 1970).

(2) Angus (Forfarshire)

The Lower Old Red Sandstone rocks in Angus and south-east Perthshire lie on the Sidlaw Anticline and the succession represents only the Arbuthnott and Garvock Groups of the Kincardineshire succession, a correlation based on palaeontological evidence found in the Arbuthnott Group sediments.

When the Arbuthnott Group is traced south-westwards along the Sidlaw Anticline major facies changes occur. The passage from the Johnshaven highland conglomerates of Kincardineshire into the Montrose Volcanic Formation has been

noted above. The olivine-basalts and andesites of this formation give rise to the high ground south of Montrose and at
Red Head These lavas in turn are succeeded by the grey-
brown sandstones, flagstones, siltstones and shales of the
Dundee Formation (2000±m), much quarried in the past and
which have yielded fish, *Climatius, Ischnacanthus, Turinea,
Cephalaspis*, the eurypterid *Pterygotus* and early land plants
including *Parka, Pachytheca, Nematophyton* and *Zosterophyllum*.
Examples of both arthropod and fish remains from these rocks
can be seen in the Dundee Museum.

The main fossil-bearing horizons in the Dundee Formation
are listed by Armstrong and Paterson (1970) and have been
used for local correlation. However, one of their findings is
that facies variation is so extensive in these rocks that the
stratigraphy set up by Hickling (1908) cannot be sustained.

The succeeding Garvock Group equivalents, by their red
colour, seen for example in the sandstones at Arbroath (Excursion 1, Location 1) and Red Head, contrast with the grey
and brown of the underlying Dundee Formation. The intervening red Auchmithie Conglomerate is a mixed conglomerate which diminishes to the south-west, being gradually
replaced by sandstones. The red cross-bedded sandstones are
often pebbly or gritty and contain calcareous detritus and
occasional calcrete soils. Their last representatives to the
south-west are conglomeratic sandstones known only from a
borehole near Tayport on the south shore of the Tay (Armstrong *et al.* 1985). There is no detailed correlation from the
south-eastern side of the Sidlaw Anticline at this level to the
north-western side around Forfar and Brechin.

(3) The Ochil Hills, Fife, Perth and the Sidlaws.

Much of the Ochil Hills falls outwith the area covered by this
guide book, but they are mentioned briefly here for complete-
ness and as an aid to correlation. This area differs from those
previously considered because volcanic rocks, the Ochil Volcanic Formation, dominate the succession. The volcanic sequence is believed to belong entirely to the Arbuthnott group

(Armstrong and Paterson 1970, p12) and is 2400m+ thick. However, the base is nowhere seen. These rocks crop out in a belt striking SW-NE along the Sidlaw Anticline. The thickness apparently decreases north-eastwards. At Bridge of Earn the Ochils lava mass splits into two ranges of hills, a more northerly one which dips north-westwards, passing through Perth to join the Sidlaw Hills where the lavas are some 1500m thick, and a southerly one dipping south-eastwards and forming the North Fife Hills.

At Bridge of Allan, at the western end of the Ochils, a fauna including *Cephalaspis, Securiaspis* and *'Pteraspis'* occurs in sediments associated with the very highest lavas. The only remaining palaeontological evidence for the age of the lavas is at Wormit in north-east Fife (Excursion 5, Location 3) where the fauna collected near the base of the lavas includes *Brachyacanthus, Ischnacanthus* and *Mesacanthus*. The entire sequence of volcanics is assigned to the Arbuthnott Group by Armstrong and Paterson (1970).

In the western Ochils there is a considerable thickness of pyroclastics and also of sediment directly derived from the erosion of the volcanic rocks. The lavas are mainly pyroxene-andesites and to a lesser extent olivine-basalts. Further eastwards at Glenfarg, the pyroclastics have thinned out leaving a predominantly andesite and basalt succession. Substantial thicknesses of both pyroclastic and volcanic conglomerates occur within the succession in the Auchtermuchty area and the volcanic pile, again predominantly andesites and basalts, is still at least 2400m thick north-west of Cupar (Armstrong *et al.* 1985, p21).

The Arbuthnott Group succession of volcanic rocks in East Fife (1800m) is well exposed in the Wormit-Tayport area (Excursion 5) and persists, thinning, north-eastwards across the Tay to Broughty Ferry. The fossil horizon at Wormit has been mentioned above. In the lavas both aphyric two-pyroxene types and feldsparphyric types are recognised. Slaggy tops and bottoms of flows are common and flow brecciation not uncommon, e.g. Excursion 7, Location 4. Between the flows volcanic conglomerates, tuffs, sandstones and pyroclastics are all present. There are abundant signs of sediment having been washed into cracks in the flows after extrusion. An upper group of lavas, at least 215m thick, follows and is exposed in quarries inland and on the shore around Tayport. The flows, which are mainly hypersthene-andesite (Geikie 1902, Pirie 1933), are 6 to 15m thick and, like the lower group, contain sandstone veins. Volcanic conglomerates are also present here. Overlying Garvock Group sandstones are known only from a borehole near Tayport (Armstrong *et al.* 1985). The succession dips south-eastwards and like that further west is unconformably overlain by Upper Old Red Sandstone sediments in Stratheden.

The Lower Old Red Sandstone rocks of the Perth area lie between the main Ochil mass to the south-west and the Sidlaws to the north-east. Eleven hundred metres of Arbuthnott Group lavas in the Ochil Volcanic Formation are exposed here, consisting mainly of basalts with a lower porphyritic group and an upper non-porphyritic group. The flows are usually 12–15m thick (Excursion 3, Location 6). Minor amounts of sediment, ranging from volcanic conglomerates to fine silts, are interbedded with the lavas and have been washed into cracks in the flows. Highland pebbles are remarkably rare. The sediments are well displayed in the roadcuts at the Friarton Interchange on the M90 motorway on the southern outskirts of Perth. They cannot be examined there, but are usually exposed in Friarton Quarry (Excursion 3, Location 6). The lavas pass north-westwards under the younger Garvock Group and Strathmore Group sediments in the Strathmore Syncline. To the south-east, faulting and younger alluvium obscure their relations with older rocks.

The Sidlaw Hills volcanics extend for 40km north-eastwards from Perth and are at first predominantly composed of olivine-basalts with minor andesites. The lavas are about 900 m thick (Harry 1956, p43), while interbedded sediments make up only 60m. As at Perth they are of Arbuthnott Group age, dip north-west at 10°–15° and pass beneath younger

Lower Old Red Sandstone sediments. The lavas thin steadily north-eastwards until in the eastern Sidlaws (Harry 1958) they make up only 60m of the succession. Basalts still predominate. The top of the volcanics is taken as the top of the Arbuthnott Group though it is known that this is a diachronous boundary (Armstrong and Paterson 1970, p11). East of Perth in the Carse of Gowrie the relations between the Sidlaw lavas and the rocks to the south-east are obscured by the North and South Tay Faults, which bring down Upper Old Red Sandstone and Carboniferous sediments into the centre of the Sidlaw Anticline, and also by Quaternary sediments along the River Tay.

The Volcanic Centres of the Lower Old Red Sandstone.

Despite the almost complete absence of vents of this age the existence of a number of centres of volcanic activity has been deduced, firstly from a study of the thickness of the lavas and pyroclastics, and secondly from a study of the distribution of volcanic conglomerates and the associated sedimentary structures which yield directions of transport.

The Highland Centre was referred to by Campbell (1913, p950) as the source of the lavas of Lower Old Red Sandstone age along the Highland Boundary Fault. No great thickness of flows is involved. They comprise some andesites and also the Lintrathen Porphyry which is an ignimbrite (Paterson and Harris 1969). A great thickness of conglomerates including volcanic conglomerates and tuffs ranging from the Stonehaven Group to the Arbuthnott Group in age has been attributed to this source (e.g. Haughton *et al.* 1990, p107). The conglomerates contain much rhyolite and acid andesite, implying the former presence of great quantities of acid volcanic rocks nearby. It was originally believed likely, by Campbell (1913), that the volcanoes lay on the other side of the Highland Boundary Fault roughly on the site of some of the Newer Granites exposed there at the present day. Modern work, e.g. Haughton (1990), Thirlwall (1989) and Trench *et al.* (1989), has done little to alter this view.

A Montrose Centre, thought by Campbell (1913, p951) to lie beneath the North Sea, was believed by him to be the source of a great thickness of andesites and basalts in the Arbuthnott Group. Local erosion of these lavas has produced volcanic conglomerates in, for example, the Dunnottar Group (Haughton, 1989).

The Ochils Centre: Armstrong and Paterson (1970, p12) reported a 2400m thick volcanic pile which interfingers with sediments including volcanic conglomerates and ashy sandstones. Agglomerates, present on the face of Dumyat, contain 'cottage-sized' blocks (Read 1927, p492). It has been suggested that the volcanic centre lies beneath the Carboniferous rocks south of the Ochil Fault.

A thick sequence including pyroclastics between Newburgh and Auchtermuchty together with volcanic conglomerates as well as flows may suggest a volcanic centre, but no vents have been described from the area. Still further eastwards 'several areas of fragmental rocks composed mainly of clasts of various types of lava... are shown, with reservations, as vents on the published map (Sheet 48E)' (Armstrong *et al.* 1985, p36), e.g. Myrecairrie Hill (Excursion 7, Location 3).

The absence of volcanic vents has been discussed by Francis (1983, pp177–81) who pointed out that in Andean volcanoes only fragments of the circumference are usually preserved, and that on the flanks of a subsiding basin these are normally asymmetrical in any case. Further, from the area of outcrop of the Old Red Sandstone volcanics, they are likely to be the products of more than one volcano. He discusses too the time available for erosion of the volcanic pile after eruption and the probability of the original edifice being completely removed and in addition the root being buried by younger sediments or volcanics. In short the chances of preservation and subsequent re-exposure of the volcanic vents from which the Lower Old Red Sandstone volcanics were erupted are not high. Fissure eruptions have also been suggested (Cameron and Stephenson 1985, p38).

As to the age of the Lower Old Red Sandstone volcanics,

Thirlwall (1981) has found problems in fitting the calc-alkaline nature of the volcanics to the plate-tectonic models proposed for the period by many workers and which he reviewed. He felt the volcanics to fit best a period of active subduction. This suggested to him that the volcanics, if they are to fit the commonly accepted tectonic model, should be of late Silurian age. Radiometric age dates discussed by him later (Thirlwall 1983, p316) also suggest a Silurian age but they are at variance with the palaeontological evidence.

Minor Intrusions

Almost confined to the Dundee and North Fife area are a considerable number of minor intrusions. They lie largely on the axial part of the Sidlaw Anticline and comprise sill-like sheets, bosses and a small number of thin dykes. Compositionally they resemble the Lower Old Red Sandstone lavas, thus the olivine-dolerites correspond to the basalts, the basic porphyrites correspond to the andesites, the porphyrites to the trachyandesites and the acid porphyrites to felsites or rhyolite.

By far the largest is the Rossie Priory Basic Porphyrite Sheet, more than 150m thick and cropping out over a distance of 8km from south-west to north-east (see Excursion 3, Location 4). Dundee Law is composed of augite-porphyrite but exposures are such that the shape of the intrusion is uncertain. Lighter coloured patches in the mass are more akin to acid porphyrite, a rock which is usually fine grained, pale coloured and feldsparphyric. One, pink in colour, an acid porphyrite, is exposed in the Ninewells railway cutting, while the pink rhyolitic felsite of Lucklaw Hill (Excursion 7, Location 1) is one of the most conspicuous landmarks in East Fife.

The chemical similarity between the lavas and the minor intrusions has led to the conclusion that they are of the same general age (Armstrong *et al.* 1985, p41), i.e. Lower Old Red Sandstone.

Conditions of sedimentation during Lower Old Red Sandstone times

The two most conspicuous features of the Lower Old Red Sandstone sediments are firstly their enormous total thickness (9km) and secondly the often huge thickness of interbedded volcanics.

Haughton (1989), from a study of the Dunnottar and Crawton conglomerates in particular, envisaged NE–SW trending basins coincident with Strathmore which were primarily controlled by sinistral NE–SW strike-slip faults. Into these a major northerly 'gravel bed river' (p524) carried much coarse material which was augmented by erosion of contemporary volcanics and, particularly during Crawton times, erosion of a metamorphic source area to the south and east, now buried beneath younger sediments within the Midland Valley. The sediments, therefore, are a result of the interplay between faulting, subsidence of the basins, vulcanicity, uplift of the sources, weathering, erosion and deposition from the rivers themselves.

The sediments thin from north-east to south-west, the Stonehaven, Dunnottar and Crawton Groups being overlapped south-westwards until at Dunkeld the Lintrathen Porphyry, an ignimbrite at the top of the Crawton Group, rests directly on Dalradian metamorphic rocks.

The Lower Old Red Sandstone volcanics increase in importance to the south-west, e.g. in the Ochil Hills. Conglomerates from these volcanics advanced to the north-west, locally in alluvial fans. Overall, however, the sandstones at many localities and at a range of horizons in Strathmore show palaeocurrent directions to the south-west. In effect Strathmore was the site of a river system draining to the south-west, bounded to the north-west by a series of alluvial fans fed from the highlands and to the south-east by volcanic uplands and a now buried metamorphic terrane (Haughton 1989).

The substantial vulcanicity during Arbuthnott times is believed by Friend and Williams (1978, p15) to have impeded drainage sufficiently to establish lacustrine conditions on a

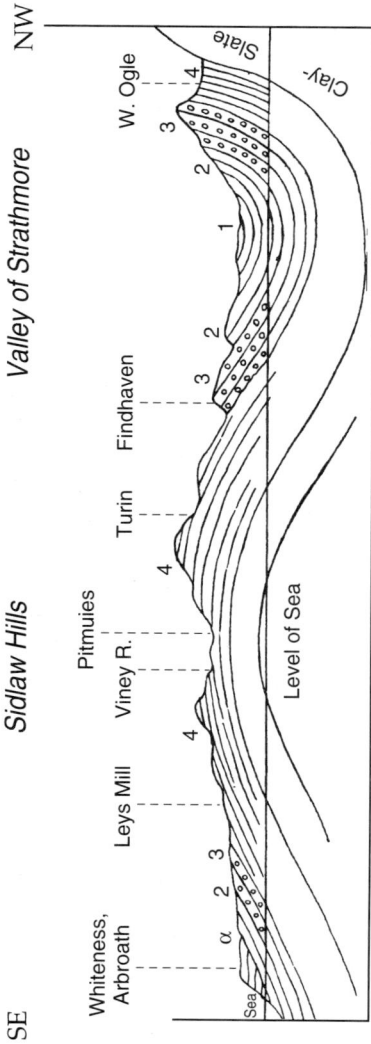

FIGURE 2: Section extending from the Highlands south-eastwards to the coast at Arbroath. After Lyell's *Elements of Geology* (1838, p.99). 1 = red marl or shale, 2 = red sandstone, 3 = conglomerate, 4 = grey paving stone, etc., α = newer deposits in horizontal beds. Most of the section comprises formations of the Lower Old Red Sandstone, but the horizontal beds at Arbroath belong to the unconformable Upper Old Red Sandstone, while the clay-slate at the NW end of the section is part of the Dalradian lying beyond the Highland Boundary Fault, not yet recognised in 1838.

number of occasions and the resulting sediments are the main fossiliferous strata. Overbank river deposits during Garvock times led to calcrete soil profiles, occasionally preserved as limestones, but more usually eroded to provide calcareous detritus. The diminution in conglomerates and increase in fine-grained sediments in Strathmore times reflect a more subdued tectonic regime.

3b. Earth movements of Middle Old Red Sandstone age

No rocks of Middle Old Red Sandstone age occur in the whole of the Midland Valley of Scotland. Rather the period was one of earth movement and erosion with little or no deposition. Two main structures in the area date from this period. These are firstly the Sidlaw Anticline, trending south-west from Montrose to Stirling where it is cut off by the Ochil Fault; on the south-east side of the Anticline the Lower Old Red Sandstone is unconformably covered by Upper Old Red Sandstone and Carboniferous rocks. Secondly, the parallel Strathmore Syncline runs south-westwards from Stonehaven on the coast and lies between the Sidlaw Anticline and the other very large structural feature of the district, the Highland Boundary Fault. The Sidlaw Anticline and the Strathmore Syncline were described early last century in some detail by Lyell (1838, pp98–101) in his *Elements of Geology* as illustrations of a typical anticline and syncline (see fig. 2).

The Sidlaw Anticline is a fairly symmetrical structure with dips of 15°–30° on either side, but the Strathmore Syncline is highly asymmetrical. The dips on the south-east limb are those of the Sidlaw Anticline, 15°–30°, but those against the Highland Boundary Fault on the north-west side are high, frequently vertical, or even overturned because of the reverse movement of the Highland Boundary Fault (Ramsay 1962), which forms the north-west flank of the Midland Valley Graben. At the time the rocks of the Highlands overrode those of the Midland Valley, dragging the rocks on the downthrow side up with them to produce a monoclinal structure.

The fold movements can be dated as Middle Old Red Sandstone, since Lower Old Red Sandstone rocks up to and including the Strathmore Group were folded and eroded before some were unconformably overlain by approximately horizontal Upper Old Red Sandstone sediments. It was estimated by Hickling (1908, p403) that in the vicinity of Arbroath 2400m of Lower Old Red Sandstone sediments were eroded away before the Upper Old Red Sandstone sediments were laid down and this figure may well be exceeded along Stratheden in Fife where the Upper Old Red Sandstone rests on Arbuthnott Group volcanics.

3c. Upper Old Red Sandstone (Excursions 1, 3, 8 and 17)

Rocks of this age are quite subordinate in amount to those of the Lower Old Red Sandstone on which they rest unconformably (Table III). In the region now under consideration they occur in three areas, listed below in order from north to south:

1. A series of small outliers, along the Kincardineshire and Angus coasts, where they are believed to be Upper Old Red Sandstone on the basis of their lithology and their position in relation to the Lower Old Red Sandstone rocks beneath.

2. An area within the Carse of Gowrie, let down by the North and South Tay Faults along the axis of the Sidlaw Anticline. These rocks have been dated by their fauna.

3. The largest area of outcrop extending south-westwards in a belt 2 to 8km wide from the mouth of the River Eden to Loch Leven and roughly coincident with the Eden Valley. These rocks have also been dated palaeontologically.

The maximum thickness, 600–900m, occurs in the Carse of Gowrie in area 2, while in area 3 it is over 600m thick around

Loch Leven. In area 1 the known thickness is not more than 60m.

1. The outliers on the Kincardine and Angus coasts are usually cut off from the Lower Old Red Sandstone by faulting, e.g. at St Cyrus and Boddin Point, respectively north and south of Montrose. In these areas the rocks comprise red, cream and sometimes mottled sandstones often showing cross bedding and channels. They are locally calcareous, sometimes sufficiently so to be identified as calcrete soils, and bear a strong lithological resemblance to the Upper Old Red Sandstone rocks of the Carse of Gowrie. At Arbroath 60m of channel and bar conglomerates and soft red sandstones crop out and a south-easterly direction of transport is indicated (Ramos and Friend 1982, p313). The pebble content corresponds closely to that of the underlying Lower Old Red Sandstone conglomerates from which they were probably derived (Excursion 1, Location 1). The base is unconformable here, the surface highly irregular and the angular discordance about 15–25°. As mentioned above, Hickling (1908) believed that somewhere in the region of 2400m of rocks were eroded away before the Upper Old Red Sandstone was laid down. This figure is probably too high for the Arbroath district, but the gap in the succession is certainly considerable.

2. In the Carse of Gowrie Upper Old Red Sandstone rocks, known as the Clashbenny Formation (600–900m, Browne 1980) have been let down by the North and South Tay Faults along the axis of the Sidlaw Anticline. A series of inliers of these rocks form the Inches or Islands of the Carse of Gowrie (e.g. at Errol and Clashbenny), gently rounded knolls with bright red soil projecting above the flat plain of the Quaternary Carse Clays. At Clashbenny, old quarries, now water filled, are famous as the type locality of *Holoptychius nobilissimus* Agassiz. The sandstones of the Clashbenny Formation are fluviatile, red in colour, and frequently have pale reduction spots. Others are more reddish brown, grey or yellow in colour and rip-up mud clasts are common. Finer grained beds

display mud cracks. These rocks are conformably overlain by the Carboniferous. However, the contact with the underlying Lower Old Red Sandstones is not now exposed.

3. Between the mouth of the Eden and Loch Leven outcrops are disappointingly few because of widespread drift deposits, the valley having been occupied by the Stratheden Glacier during the Pleistocene glaciation. There are, however, two main areas of outcrop, the first of which lies around the West Lomond and Bishop Hill (Excursion 17, Locations 1–3). Only here can the contact between the Upper Old Red Sandstone, approximately 600m thick, and the overlying Carboniferous rocks be seen. The second area lies around Cupar and in it the thickness is approximately 300m. The best exposures are in the famous Dura Den section 5km east of Cupar (Excursion 8). Quarry exposures are not important except that at Drumdryan, Cupar, which is of historical interest as the first locality from which *Holoptychius* scales were obtained by Fleming in 1831.

In the Stratheden–Lomond Hills area Chisholm and Dean (1974) subdivided the Upper Old Red Sandstone into a series of formations, now taken together as the Stratheden Group (Paterson and Hall 1986), the two lowest, the Burnside and Glenvale, comprising red to yellow, fine to coarse-grained and pebbly sandstones. Intra-formational clasts are common and trough cross bedding indicates an eastwards flow. The overlying Knox Pulpit Formation is of sandstones, some with millet-seed grains and showing cross bedding and vertical burrows among other features, now interpreted as wind-blown dune deposits (Hall and Chisholm 1987). The formation does not extend west of Kinross. The rather local Dura Den Formation (30m) (Excursion 8, Locations 8, 9) is famous for its fossil fish fauna including *Bothriolepis, Phyllolepis, Glyptopomus, Eusthenopteron, Holoptychius* and *Phaneropleuron* of Fammenian age which occurs in cream coloured sandstones alternating with red, cream and green siltstones, some with mudcracks.

References

ARMSTRONG, M. and PATERSON, I. B., 1970. The Lower Old Red Sandstone of the Strathmore Region. *Rep. Inst. Geol. Sci.* No. 70/12.

ARMSTRONG, M., PATERSON, I. B., and BROWN, M.A.E., 1985. Geology of the Perth and Dundee district. *Mem. Br. Geol. Surv.,* Sheets 48W, 48E, 49.

BROWNE, M. A. E., 1980. The Upper Devonian and Lower Carboniferous (Dinantian) of the Firth of Tay, Scotland. *Rep. Inst. Geol. Sci.* No. 80/9.

CAMERON, I. B. and STEPHENSON, D., 1985. The Midland Valley of Scotland. *Brit. Reg. Geol.* 3rd Ed.

CAMPBELL, R., 1913. The geology of south-eastern Kincardineshire. *Trans. Roy. Soc. Edinb.,* **48**, 923–60.

CHISHOLM, J. I. and DEAN, J. M., 1974. The Upper Old Red Sandstone of Fife and Kinross: a fluviatile sequence with evidence of marine incursion. *Scot. Jour. Geol.,* **10**, 1–30.

FRANCIS, E. H., 1983. Magma and sediment – II. Problems of interpreting palaeovolcanics buried in the stratigraphic column. *Jour. Geol. Soc. Lond.,* **140**, 165–83.

FRIEND, P. F. and WILLIAMS, B. P. J., 1978. International symposium on the Devonian System (PADS 78). A field guide to selected outcrop areas of the Devonian of Scotland, the Welsh borderland and South Wales. *The Palaeontological Association.*

GEIKIE, A., 1902. The geology of Eastern Fife. *Mem. Geol. Surv. Scotld.*

HALL, I.H.S. and CHISHOLM, J.I., 1987. Aeolian sediments in the late Devonian of the Scottish Midland Valley. *Scot. Jour. Geol.,* **23**, 203–8.

HARRY, W. T., 1956. The Old Red Sandstone lavas of the western Sidlaw Hills, Perthshire. *Geol. Mag.,* **93**, 43–56.

—————, 1958. The Old Red Sandstone lavas of the eastern Sidlaws. *Trans Edinb. Geol. Soc.,* **17**, 105–12.

HAUGHTON, P. D. W., 1988. A cryptic Caledonian flysch terrane in Scotland *Jour. Geol. Soc. Lond.,* **145**, 685–703.

—————, 1989. Structure of some Lower Old Red Sandstone conglomerates, Kincardineshire, Scotland: deposition from late-orogenic antecedent streams. *Jour. Geol. Soc. Lond.* **146,** 509–25.

——————, ROGERS, G. and HALLIDAY, A. N., 1990. Provenance of Lower Old Red Sandstone conglomerates, SE Kincardineshire: evidence for the timing of Caledonian terrane accretion in central Scotland. *Jour. Geol. Soc. Lond.*, **147**, 105–20.

HICKLING, G., 1908. The Old Red Sandstone of Forfarshire. *Geol. Mag.* Dec. 5, vol. 5, 396–408.

HOUSE, M. R., RICHARDSON, J. B., CHALONER, W. G., ALLEN, J. R. L., HOLLAND, C. H., and WESTOLL, T. S., 1970. A correlation of the Devonian rocks in the British Isles. *Spec. Rep. Geol. Soc. Lond.* No.7.

LYELL, C., 1838. *Elements of geology.* Murray, London.

MYKURA, W., 1991. The Old Red Sandstone. In Craig, G. Y. (ed) *The Geology of Scotland.* Geological Society, London, 297–344.

PATERSON, I. B. and HALL, I. H. S., 1986. Lithostratigraphy of the late Devonian and early Carboniferous rocks in the Midland Valley of Scotland. *Rep. Br. Geol. Surv.* **18**, No.3.

——————— and HARRIS, A. L., 1969. Lower Old Red Sandstone ignimbrites from Dunkeld, Perthshire. *Rep. Inst. Geol. Sci.* No. 69/7.

PIRIE, H., 1933. The petrography of the Lower Old Red Sandstone lavas in the neighbourhood of Tayport. *Trans. Proc. Perthsh. Soc. Nat. Sci.*, **9**, 97–106.

RAMOS, A. and FRIEND, P. F., 1982. Upper Old Red Sandstone sedimentation near the unconformity at Arbroath. *Scot. J. Geol.* **18**, 297–315.

RAMSAY, D. M., 1962. The Highland Boundary Fault: reverse or wrench fault? *Nature, Lond.* **195**, 1190.

RAYNER, D. H. 1981. *The stratigraphy of the British Isles.* (2nd Ed.) Cambridge.

READ, H. H., 1927. The western Ochil Hills. *Proc. Geol. Ass. Lond.*, **38**, 492–4.

THIRLWALL, M. F., 1981. Implications for Caledonian plate tectonic models of chemical data from volcanic rocks of the British Old Red Sandstone. *Jour. Geol. Soc. Lond.*, **138**, 123–38.

——————, 1983. Discussion on implications for Caledonian plate tectonic models of chemical data from volcanic rocks of the British Old Red Sandstone. *Jour. Geol. Soc. Lond.*, **140**, 315–8.

——————, 1989. Movement on proposed terrane boundaries in

northern Britain: constraints from Ordovician-Devonian igneous rocks. *Jour. Geol. Soc. Lond.*, **146**, 373–6.

TRENCH, A., DENTITH, M. C., BLUCK, B. J., WATTS, D. R. and FLOYD, J D., 1989. Palaeomagnetic constraints on the geological terrane models of the Scottish Caledonides. *Jour. Geol. Soc. Lond.*, **146**, 405–8.

Chapter 4

Carboniferous

Tables IV, V and VI
Excursions 8 – 18

Outcrops of rocks belonging to the Carboniferous System lie mainly south-east of a line extending from Guardbridge, at the mouth of the River Eden, to the south side of Loch Leven, except for the Lomond Hills mass, which projects to the north-west, and small faulted outliers occurring near Errol, Newburgh and Bridge of Earn (see Map 2). Within this area all the major units of the Carboniferous System are represented. The outcrop distribution is partly controlled by two large complementary structures trending N–S: (1) the Leven Syncline in which the Upper Carboniferous is preserved and which continues across the Firth of Forth to join the Lothian Syncline, and (2) the Burntisland Anticline in which Lower Carboniferous rocks are exposed. The Carboniferous rocks are subdivided as shown in Table IV. They are described in ascending order from bottom to top, volcanic rocks being included in their correct stratigraphical position.

Inverclyde Group

Rocks of this age are present (1) at Fife Ness and Kingsbarns (2) in Stratheden and at Loch Leven and (3) in the Carse of Gowrie.

1. At Fife Ness and Kingsbarns (Excursion 11) the Balcomie Beds consist of yellow and reddish sandstones with green and

TABLE IV: Correlation table for the Carboniferous of Fife

Sub-system (after Francis 1991)	Series (Francis 1991)	Stage (after Browne 1986 & Francis 1991)	Old Classification MacGregor (1960)		New Classification Paterson & Hall (1986), Chisolm et al (1989)	Formations
SILESIAN (300my / 310my)	Westphalian	D / C	Upper Coal Measures			
		B	Middle Coal Measures — Skipsey's MB, Queenslie MB			
		A	Lower Coal Measures			
	Namurian		Passage Group		Passage Formation	
			Upper Limestone Gp — Castlecary Lst, Index Lst		Upper Limestone Formation	
			Limestone Coal Group		Limestone Coal Formation	
(325my)		Brigantian	Lower Limestone Gp — Top Hosie Lst, Hurlet Lst		Lower Limestone Formation	St Monans Brecciated Lst
DINANTIAN	Viséan	Asbian	Upper Oil-shale Group	Calciferous	Strathclyde Group	Pathhead; Sandycraig; Pittenweem; Anstruther; Fife Ness
		Holkerian / Arundian / Chadian	Lower Oil-shale Group	Sandstone		
			Cement-stone Group	Measures	Inverclyde Group	Balcomie; Ballagan; Kinnesswood
	Tournaisian	Courceyan	Upper Old Red Sandstone		Stratheden Group	Knox Pulpit; Dura Den } (Clash-bennie); Glenvale; Burnside
DEV-ONIAN (355my)	Fammenian					

41

purple marls and rare purple conglomerates with andesite pebbles up to 2cm long in a muddy matrix. Calcrete soils are developed. No fossils have been found in these rocks but they resemble the Inverclyde Group Downie's Loup Sandstone of the Stirling district (Browne 1980a). The andesite pebbles imply not too distant erosion of Lower Old Red Sandstone lavas at the time of deposition.

2. In Stratheden and around the Lomond Hills the Kinnesswood Formation also includes calcrete soil profiles in a sequence of pebbly, fluviatile sandstones. Although absent from Bishop Hill (Excursion 17) Ballagan Formation cementstones and clays are present at Kingskettle (130m: Browne 1980a, p325), Benarty and Navitie.

3. In the Carse of Gowrie the Kinnesswood Formation (33m: Browne 1980b, p4) is known mainly from boreholes. There too it comprises reddish and yellow sandstones with calcrete soils and mudstones. The Ballagan Formation (240m+) comprises mudstones with cementstones and some sandstones. Salt pseudomorphs, desiccation cracks and a restricted fauna point to lagoonal conditions of formation while plant spores confirm a Tournaisian age for the rocks.

Strathclyde Group

In the Burntisland area the oldest Strathclyde Group rocks belong to the upper part of the Calders Member of the West Lothian Oil–Shale Formation just below the horizon of the Pumpherston Shell Bed (see Table V), a marine horizon carrying the goniatite *Beyrichoceratoides*. One hundred metres above it the Burdiehouse Limestone, a fresh-water limestone 6m thick, marks the base of the Hopetoun Member. Forty-six metres above the limestone is the 2m-thick Dunnet Oil Shale, worked a century ago and the only noteworthy representative of the thick oil shales of the Lothians to occur in the area. The upper part of the Hopetoun Member is replaced by 430m+ of volcanic rocks, mainly basalts with some tuffs (Francis 1991b,

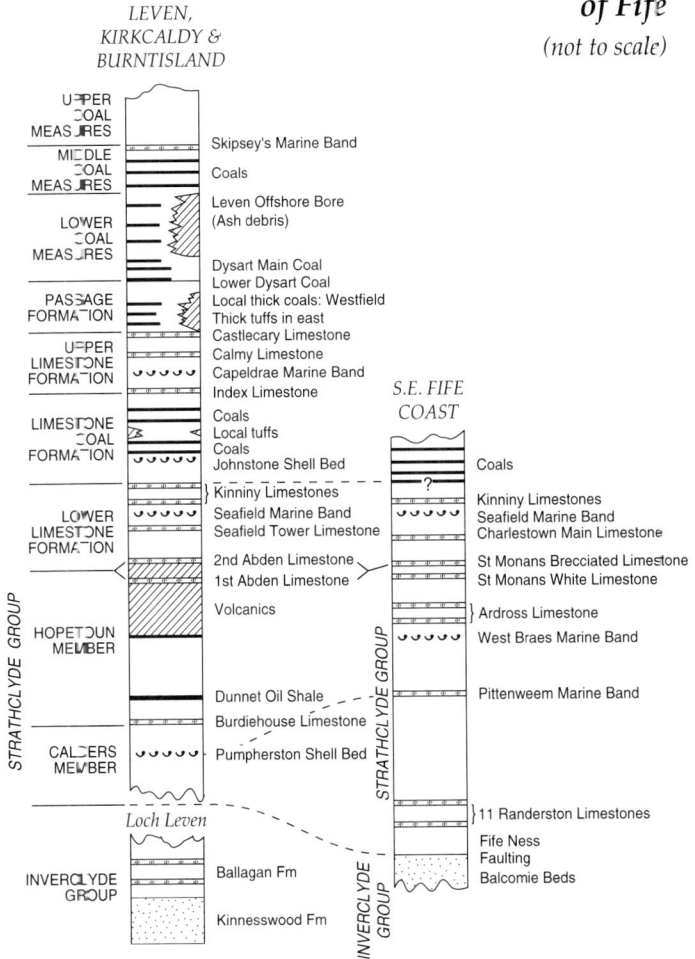

Carboniferous Rocks of Fife
(not to scale)

TABLE V: **Carboniferous rocks of Fife.**

p399). A number of vents in the vicinity of Burntisland are the most likely source of these lavas.

The Burntisland succession can be correlated with the much thicker Lothian succession to the south-west by means of the Pumpherston Shell Bed, the Burdiehouse Limestone and the Dunnet Oil Shale and it seems that the Burntisland district lay near the eastern edge of a large basin of lacustrine sedimentation in which the oil shales were laid down (George 1958). The Pumpherston Shell Bed, equivalent to the Macgregor Marine Bands (Wilson 1974), is the only horizon which allows correlation of the West Lothian Oil–Shale Formation sequence with the Strathclyde Group rocks of East Fife, palaeontological zonation with spores still being incomplete.

The Strathclyde Group attains its greatest thickness in East Fife in the Anstruther Anticline where the succession is over 2000m thick without reaching the Inverclyde Group believed to lie beneath (Browne 1980a, p326). The rock types in order of abundance are (1) sandstones, often with ripple marking and cross bedding; (2) clays and shales often with nodules of clayband ironstone; (3) seatearths (fossil soils), though rarely with coal over them, those that do occur usually being very thin; (4) limestones, which are very rare and seldom more than 30cm thick. A cyclic repetition of these rock types is characteristic (Francis 1991a, p347) and much research has been devoted to a study of this, e.g. Fielding *et al.* (1988).

Some of the limestones, which are usually dolomitised, and a few of the shales are marine and may carry crinoid ossicles, gastropods, the brachiopods *Lingula, 'Productus', Schizophoria,* and *Rhynchonella (Camarotoechia)* and the bivalves *Naiadites* and *Schizodus.* The St Monans White Limestone yielding the corals *Lithostrotion* and *Dibunophyllum* is unique in the area. These marine horizons are generally impersistent and only the Pittenweem Marine Band, equivalent to the Macgregor Marine Bands of East Lothian (Wilson 1974) (Excursions 9 and 10) has been detected over any considerable part of the district. The non-marine shales yield bivalves and ostracods and usually finely comminuted plant debris, while the sandstones commonly contain *Stigmaria* (roots) and stems of *Lepidodendron*.

The formations of the Strathclyde Group, defined on their lithology, have been traced over East Fife by Forsyth and Chisholm (1977, Fig. 2, p12) and there is much detailed information in their memoir.

The Fife Ness Beds (230m+), the lowest formation of the Strathclyde Group, are predominantly thick white sandstones with interbedded grey and sometimes red mudstones. Marine horizons are unknown but both algal and ostracod-bearing carbonates occur.

The Anstruther to Pathhead, St Monans, coast section exposes some 1600m of Strathclyde Group rocks, extending through the four higher formations. Taken with the succession in the Anstruther Bore (300m: Forsyth and Chisholm 1977, p22) the Anstruther Beds are 800m+ thick and are conspicuously cyclic with thin dolomitic carbonates, some of which are marine, passing up into shales, siltstones and sandstones with some seatearths and thin coals. These beds are well displayed at Randerston (Excursion 11).

The Pittenweem Beds (200m+), though less clearly cyclic, also comprise shales, siltstones and sandstones. Dolomites are rare but the shales or mudstones not infrequently include ironstones. Among the marine beds the Pittenweem Marine Band (6m), correlated with the Macgregor Marine Bands in East Lothian and the Pumpherston Shell Bed at Burntisland, contains crinoid ossicles and a marine brachiopod fauna.

Between Pittenweem Harbour and Swimming Pool typical Sandy Craig Beds (550m+) consist of both grey and red shales between thick sandstones, some of which are very coarse grained. Only one marine band is known and the red mudstones include concretionary carbonates (calcrete soils) some of which have been reworked, e.g. that exposed at Pittenweem Swimming Pool.

The Pathhead Beds (311m), the highest formation of the Strathclyde Group, crop out on the coast westwards from Pittenweem Swimming Pool to Pathhead (Excursion 12). They consist of cyclic sediments which show not only increasingly marine horizons towards the top, e.g. the Ardross and St Monans White Limestones, but also thin coals and seatearths, for example those beneath the Ardross Limestones.

It appears that the 2000m+ thick succession of the Anstruther area has thinned to about 490m at St Andrews, i.e. northwards. Furthermore, when the succession is traced south-westwards from there it diminishes to 90m at Cults, south-west of Cupar, and to only 30m on Bishop Hill overlooking Loch Leven. It is believed that the Burntisland Anticline towards which this thinning takes place was not so much an area of uplift, but rather one of much less subsidence than the ground on either side during Strathclyde Group times; indeed spores suggest that only the very highest Strathclyde Group beds are present there (Browne 1980a, p324). Further west in a thin Strathclyde Group succession in the Cleish Hills minor volcanics are the 'feather edge' of the Clyde Plateau Volcanic Formation.

Conditions of Sedimentation

The nature of the sediments points to their deposition under humid, tropical, deltaic, fluviatile, swamp and occasionally marine conditions with forests being killed off repeatedly by subsidence and drowning (Belt 1975). Channel switching within the delta, against a background of general subsidence, and fluctuations in sea level, are also believed to play a part in the formation of the cyclic sediments of the Carboniferous. Flooding and truly marine conditions were rare and short lived, e.g. some of the Randerston Limestones, the Pittenweem Marine Band, the Ardross Limestones and the St Monans White Limestone. In the west only the Pumpherston Shell Bed is marine and this suggests that open sea lay to the east at this time (Greensmith 1965, p241).

Lower Limestone Formation

Sediments belonging to this formation overlie the Strathclyde

Group and are exposed on the coast at St Monans and Elie in the east and Kinghorn in the west. Inland many quarries have been opened up in the limestones, but few are now worked. Good exposures can still be seen at Cults, south-west of Cupar, on Bishop Hill (Excursion 17), the East Lomond (Excursion 16), at Charlestown and at Roscobie, 5km north of Dunfermline. Structurally the group outcrops on either side of the Burntisland Anticline and round the southward plunging Leven Syncline.

Lithologically there is a marked similarity between the cyclic sediments of this formation and those of the Strathclyde Group in East Fife, particularly the highest Pathhead Beds. However, more of the Lower Limestone Formation comprises marine shales and limestones and the latter are thicker and more persistent such that they are used for correlation over much larger areas (Wilson 1989). It is customary to take the lowest of the limestones as equivalent to the Hurlet Limestone of the west of Scotland. The correlation of this and other limestones higher in the group is set out in Table VI.

The Kinghorn succession is about 146m thick and roughly half of it is marine. Twenty-four metres of lavas and tuff overlie the First Abden Limestone, but do not occur anywhere else in the area at this horizon. Westwards the group thickens to 190m around Dunfermline, on the west side of the Burntisland Anticline, due mainly to an increase in the thickness of the shales. To the east in the St Monans Syncline the sequence, completely exposed except for the top few metres, is 180m thick.

Conditions of sedimentation

The area underwent subsidence with intermittent flooding by the sea such that clear-water marine limestone formed. On each occasion the area was then silted up with marine clays followed by non-marine clays which gave way in turn to bedded and then cross-bedded sandstones as deltaic conditions spread across the area. These sandstones are frequently covered by seatearths (fossil soils) and coal seams indicating

Standard names of Wilson (1989)	WEST of SCOTLAND	Limestones of the Lower Limestone Group			
		WEST FIFE	KINGHORN-KIRKCALDY	BISHOP HILL	ST MONANS
	Four Hosie Limestones	Kinniny Limestones & Seafield MB	Kinniny Limestones & Seafield MB	—	Kinniny Limestones & Seafield MB
	Blackhall	Charlestown Main	Seafield Tower	Charlestown Main	Charlestown Main
	Craigenhill	Charlestown Green	—	—	St Monans Little
	Hurlet	Charlestown Station	Abden Limestones	Charlestown Station	St Monans Brecciated

TABLE VI: Limestones of the Lower Limestone Group in Fife and their correlation.

that forests grew on top of the deltas. Gentle subsidence during the period of forest growth was followed by more rapid tectonic subsidence (Fielding *et al.* 1988, p252) and a fresh invasion of the sea, giving rise to marine clays and then clear-water marine limestone once more.

Limestone Coal Formation

This formation has been of great economic importance in the area since it contains the lower set of workable coals of the Fife coalfields. Formerly extensively worked in the Cowdenbeath and Lochgelly areas, old workings are also to be found around Kirkcaldy and Dunfermline and to a lesser extent at Elie, St Monans, Colinsburgh, Largoward, Denhead and Ceres. Structural complexities and the presence of numerous igneous intrusions were responsible for the closing down of pits in most of the small fields.

The very large number of workings has produced a great deal of information about the Limestone Coal Formation, but surface exposures are scattered and no good section is available for study in the area under review. The formation is delimited by the Upper Kinniny Limestone at the base and by the Index Limestone at the top. Only two widespread marine horizons occur between these in Fife. The lower, the Johnstone Shell Bed (see Table V) is well known and carries *Lingula*; the other, the Black Metals, is rarely seen. The greater part of the succession is made up of thin-bedded micaceous sandstones and cross-bedded sandstones, with smaller thicknesses of siltstones, shales and clays with clayband ironstones. Seatearths and up to 17 workable coals account for an even smaller part of the succession. During the 19th century ironstones were worked around Cowdenbeath and Lochgelly and to a lesser extent elsewhere in Fife including Denhead, near St Andrews.

The total thickness of the formation varies considerably at different localities with the line of the Burntisland Anticline again clearly one of minimum subsidence.

Variations in thickness in the Limestone Coal Formation:

Cowdenbeath	Burntisland Anticline	Kirkcaldy
430m total	155m total	259m total
30m of coal	6m of coal	15m of coal

Except in the Saline district volcanic activity was very minor during the period, giving rise to only a few metres of tuffs.

Conditions of sedimentation.

These were similar to those of the Lower Limestone Formation, but forested deltaic swamp conditions predominated and led to the prolonged accumulation of peats, now seen as thick coals. Subsidence varied over the area, but only twice were marine conditions introduced and even these prevailed only for brief periods.

Upper Limestone Formation

The formation crops out in two main areas, unfortunately largely obscured by drift. The first of these, the Cowdenbeath Syncline, lies on the west side of the Burntisland Anticline. From there a narrow outcrop extends north-eastwards to Westfield, before crossing the Burntisland Anticline and turning south to the coast at the north-east end of Kirkcaldy. Until the 1950s it was well exposed there, but dumping from the pits on the adjacent coast to the north-east had almost entirely obscured the section by 1962. The second area of outcrop is north and north-west of Lundin Links, but exposure is poor and most of the information has been obtained from boreholes.

The base of the formation is drawn at the horizon of the Index Limestone. Above it the Upper Limestone Formation varies in thickness as follows:

Cowdenbeath-Lochore Syncline	Burntisland Anticline	North of Lundin Links
335m est.	238m	320m

The Index Limestone (see Table IV) is represented in Central Fife by a marine shale. Similarly the next two limestones of the standard succession of the west of Scotland, the Lyoncross and Orchard Limestones, are represented in central Fife by calcareous marine shales named respectively the Lochore and Capeldrae Marine Bands. Only on reaching the metre-thick Calmy Limestone can a lithological correlation be made between Fife and the west. The Plean Limestones of the Scottish Central Coalfield are also represented in Fife by calcareous shales. The Castlecary Limestone at the top of the formation, and still exposed on the coast north-east of Kirkcaldy, is about 2m thick in Central Fife and is associated with calcareous shales. There are many fireclays or seatearths, but only a small number of coal seams. However among these the Hirst Coal in West Fife and Clackmannanshire beyond is being extensively worked for the huge Longannet coal-fired power station. The sediments are similar to those of the Limestone Coal Formation, the main difference in Fife being a small increase in the proportion of marine beds. Contemporary volcanic activity was considerable in the area north of Leven and it is believed that much of the tuff there came from vents in the vicinity of Largo Law (Forsyth and Chisholm 1977, p105). At Westfield there are both tuffs and five basalt flows of this age.

Conditions of sedimentation

The same type of cyclical sedimentation as was seen in the Lower Limestone Formation occurs in the Upper Limestone Formation, the main difference being that clear-water limestone formation but rarely took place. As a result the fauna of the marine beds is usually impoverished.

Passage Formation

Rocks belonging to this formation crop out in two areas, a small one at Westfield in the north-east-trending Cowden-beath Syncline and terminating against the Ochil Fault, and a much larger one bounding the Leven Syncline on its north and west sides. This larger outcrop runs northwards from the coast near Kirkcaldy, and then north-eastwards for some distance towards Largo Law, with a branch occupying the centre of the Earl's Seat Anticline. As with the previous groups the Passage Formation is quite thin – 137m – adjacent to the Burntisland Anticline, whereas to the east it is over 270m thick.

In the area north-east of Kirkcaldy and in the Earl's Seat Anticline the bulk of the succession comprises pale-grey or white cross-bedded sandstones and grits with subordinate seatearths and shales, often red or purplish in colour. Marine horizons are rare and limestones absent. The base of the succession is drawn at the Castlecary Limestone and its top is arbitrarily drawn at the base of the Lower Dysart Coal. On the coast between Kirkcaldy and Dysart there are substantial exposures of sandstones, often very coarse grained, but virtually nothing else. Palaeontologically the top boundary lies well below the Lower Dysart Coal (Francis 1991a, p376). When this standard succession is traced eastwards, first a single basalt lava flow appears 3km north of Leven and then as Largo Law is approached most of the top half of the group is replaced by tuffs from that vent.

A major change in the succession also takes place westwards. Not only does the formation thin to about half the previous thickness, but in the Westfield Basin coal seams account for 60m of the 150m of strata present (Francis 1991a, p364). These coals were apparently very restricted in distribution, occurring only at Westfield where they were worked opencast for some twenty years, yielding some 21 million tonnes of coal.

Conditions of sedimentation

Deltaic corditions again prevailed, together with fluviatile, with few marine invasions. The cross-bedded sandstones and grits often show slumped foreset beds and frequent seatearths, and the thick coals at Westfield indicate that forests were repeatedly established on the delta, sometimes for prolonged periods. Thickness variations are again due to differential subsidence.

Coal Measures

The outcrop of these rocks is restricted to a south-eastward-dipping faulted strip extending along the coast from Dysart to Leven, a synclinal strip extending 8km northwards from Dysart and, formerly, a small downfaulted area at Westfield, now virtually removed by opencast mining. At one time almost the whole succession was exposed between Dysart and Leven but dumping from pits on the coast in the 1960s largely obscured this. The succession thins north-eastwards from nearly 800m at Dysart to 580m at Leven (Knox 1954) and this thirning is known, on the basis of much detailed knowledge from mining, to affect both individual coal seams and the strata between.

The Coal Measures are subdivided as follows (see Table IV):

Upper Coal Measures	Westphalian C and D	(= Barren Red Measures)*
Middle Coal Measures	Westphalian B	(= Productive Measures)*
Lower Coal Measures	Westphalian A	

* = Pre–1960 terminology

Skipsey's Marine Band separates the Middle from the Upper Coal Measures and the Queenslie Marine Band separates the Lower from the Middle Coal Measures. Apart from these two

53

marine bands and rare local *Lingula* bands the strata are entirely non-marine, and zonation and correlation are therefore based on the distribution of fresh-water bivalves, or 'mussels' as they are usually called, or on plant spores. Both groups of fossils have been used to establish detailed zonal schemes in the Westphalian and both show that the base of the Westphalian lies some distance below the convenient base of the Coal Measures drawn in Fife at the bottom of the Lower Dysart Coal (Francis 1991a, pp365 and 376), the lowest workable seam.

In the Lower and Middle Coal Measures a maximum of twenty coal seams have been worked, the number falling off to the north-east. Many of the seams have partings of shale or fireclay and a few have partings of cannel coal or ironstone. The beds between the coals include white and buff sandstones, often cross-bedded, and dark grey shales with plant remains and occasional sandy partings. Mussel bands are usually found in the roofs of coal seams, but may occur in the shales as well. This type of succession shows little lateral variation except in thickness and extends as high as Skipsey's Marine Band which contains *Lingula*, the nautiloid *Metacoceras*, the goniatite *Anthracoceras* and fish remains, all of which were at one time collected on the shore at Wemyss Castle.

The Upper (Barren Red) Coal Measures make up the rest of the succession. The strata consist of sandstones and grits, shales, clays and marls, almost all with a red or purple colour indicating a semi-arid climate.

Occasional thin beds of probably fresh-water limestone are present together with a few thin coals. Plant fossils have been recorded as well as *Spirorbis*, fish remains and Crustacea (Binney and Kirkby 1882), but fossiliferous exposures are not now known. The beds attain a maximum thickness of 300m without the top being seen.

Only a single tuff band one metre thick, lying above the Lower Dysart Coal, has been seen at outcrop in the whole of the Coal Measures. It is all the more surprising, therefore, to find that in the offshore bore at Leven, after penetrating the

Upper and Middle Coal Measures and the upper part of the Lower Coal Measures, the bore passed through an 8m basalt flow followed by 150m of 'volcanic detritus' in which tuffs, volcaniclastic sediment and even thin coals occur and in which the bore terminated (Ewing and Francis 1960). The bore log suggests that a volcanic centre in this vicinity was undergoing erosion and renewal at this time. The Lundin Links Neck cuts Middle Coal Measures sediments and the Viewforth Neck, Lower Largo, has in it collapsed tuffs containing Lower Coal Measures spores (Forsyth and Chisholm 1977, pp180–3).

Carboniferous Earth Movements

The effects of earth movements during and near the end of the Carboniferous Period can be clearly seen in Fife, but further north they are much more difficult to assess owing to the scarcity of post-Middle Devonian sediments. The movements can be divided into two categories, those contemporary with Carboniferous sedimentation and those that are clearly later.

(i) The movements contemporary with sedimentation can be most readily discerned in a strip, approximately 20km wide from east to west, lying north of Burntisland and extending northwards to the Lomond Hills. This strip includes the Burntisland Anticline which persisted as an area of minimum subsidence from early Strathclyde Group times to Passage Formation times. Details of the thinning of each group of sediments over the anticline are given above and the total effect is summarized below:

Cowdenbeath area	Burntisland Anticline	Kirkcaldy area
1128m+	674m	1963m

This variation across the anticline affected not only individual beds but also whole groups of beds. A basement control has been invoked by Francis (1991a, p376). A similar but apparently unrelated variation across the Earl's Seat Anticline is confined to the Lower and Middle Coal Measures, this anticline having had no apparent effect on sedimentation prior to this time.

Synsedimentary movements on faults have been demonstrated on for example the Ochil Fault at Westfield (Brand *et al.* 1980, p6) while a close relationship between the Ardross Fault and volcanic activity has been deduced by Francis and Hopgood (1970).

(ii) The late-Carboniferous, or Hercynian, earth movements gave rise to several folds of considerable amplitude, to many minor folds and to widespread faulting, locally with large displacements.

From west to east the major folds in Fife attributed to the late-Carboniferous movements are: (1) the accentuation of the Burntisland Anticline and the complementary syncline to the west between Cowdenbeath and Westfield; (2) the main N-S trending Leven Syncline which extends across the Forth to the Lothian Syncline near Edinburgh and containing the Coal Measures, but subdivided by subsidiary anticlines at Earl's Seat and Leven; (3) a large and irregular syncline extending to the north and north-east from Largo almost to St Andrews; (4) the St Monans Syncline and the Anstruther Anticline.

A host of minor folds can be seen wherever exposure permits, e.g. those along the Kinkell Braes east of St Andrews (Excursions 9 and 10). Evidence for the existence of numerous small folds and some not so small can also be obtained from mining data, e.g. in the Ceres Coalfield in which the strata reach a vertical position.

North of the Carboniferous outcrop the folding was apparently much less severe, the Upper Old Red Sandstone along Stratheden being almost horizontal except in the vicinity of faults (Excursion 8).

To the south of Loch Leven the largest fault in the area, in common with many minor ones, strikes east–west and is a continuation of the great Ochil Fault of the Stirling area. Its southerly downthrow is at least 750m, much of it post-Coal

Measures, although some of these faults are now known to have been active by Namurian times (Francis and Walker 1987). Eastwards the fault splits up, but the individual branches are still important at Leven and Durie. Large numbers of parallel faults are known from mining around Dunfermline, Cowdenbeath and Glenrothes (e.g. Francis 1961, p35).

Post-Dinantian faulting was responsible for the preservation of the Carboniferous rocks at Bridge of Earn, Newburgh and Errol (Armstrong *et al.* 1985, pp62–4) and, by inference, outliers of Upper Old Red Sandstone elsewhere in Angus and Kincardineshire.

Carboniferous Intrusive Igneous Rocks

In the area covered by the guide there are perhaps seventy sills, many of which form parts of the major quartz-dolerite Midland Valley Sill (Francis 1991b, p407), about a dozen major dykes and about a hundred volcanic vents. Some at least of the distribution is clearly related to basement structure, e.g. the vents along the Ardross Fault and others adjacent to lines of reduced subsidence in the Carboniferous (Francis 1991b, p410). It was clearly an area of very considerable igneous activity and the products can be divided into three categories.

1. A suite of alkali-dolerite sills in the Namurian rocks of Fife; these are believed by Francis and Walker (1987) to be a little younger than the sediments in which they lie, sediments which at the time of emplacement must have been almost unconsolidated and still water saturated and thus of low density compared with the dolerite magma. Spatially and petrochemically they are closely related to volcanic necks which Francis and Walker believe were the feeders for the sills. They are composed variously of basic plagioclase feldspar, clinopyroxene, olivine, analcite and amphibole.

2. The quartz-dolerite intrusions (Excursions 3, 8, 16 and 17)

comprise sills and E–W dykes and are composed of basic plagioclase feldspar and pyroxenes with interstitial quartz and micropegmatite. Pyrite is a common accessory mineral and hypersthene has also been recorded. Where the interstitial material appears to be devitrified glass the rocks are described as tholeiites (Excursion 3). Age dating at 296 My suggests a very late Carboniferous (Stephanian) age for these (Francis 1991b, p407). The sills appear to have been fed by E–W dykes. From these the magma seems to have spread out as sills down-dip into adjacent thick sedimentary basins (Francis 1982). The E–W quartz-dolerite and tholeiite dykes of North Fife, Perthshire, Angus and Kincardineshire are grouped with those of the rest of the Midland Valley on the basis of similarity of trend and the uniformity of composition (Francis 1991a, p380).

3. Of the one hundred or so volcanic vents in East Fife a number have within them basanite intrusions. Age dating on a number of these, e.g. Elie Ness (Forsyth and Rundle, 1978), points to around 289±10 My as the likely age, thus suggesting a late phase of post-quartz-dolerite igneous activity. Unlike Ayrshire there is no evidence of sill emplacement of this early Permian age.

Intrusion Mechanism of the Carboniferous-Permian Vents

Considering the nature of these vents two points are clear: (1) where the magma carried xenoliths of ultrabasic rocks from the mantle it must have come from considerable depths, and (2) evidence of any volcanic edifice still in place is lacking, but is perhaps hardly to be expected (Francis 1983, p175). In most vents the present level of exposure is many hundreds of metres below what was the surface at the time of eruption (Forsyth and Chisholm 1977, p175), e.g. Excursion 13.

It is evident that neck drilling from below has been brought about by gas fluxion breaking down the country rock to produce both a breccia of country rock fragments, e.g. sandstone and shale (Excursions 10 and 15), and a finer-grained equivalent, tuffisite, which invades the wall rock as thin dykes and stringers. Similar material fills the space between the larger clasts within the vents. Extensive wall-rock stoping and, later, collapse into the vent took place during each pulse of activity, the site marked by inward dipping concentric fractures and steep concentric dips in the bedded tuffs (Excursions 10 and 15). In many vents a progressive increase in content of igneous material inwards from the vent margin can be seen with the appearance in the tuffs of shreds of then fluid but now highly altered and whitened basalt, or fragments of already solidified basalt ranging from columnar-jointed blocks down to the finest dust. Agglomerates and tuffs, greenish in colour and comprising predominantly igneous material, contrast with others, almost entirely composed of sediment, which have a grey hue. Mixtures are not uncommon.

Centroclinally dipping bedded tuff is common in the larger vents, e.g. Kincraig (Excursion 15) or the Rock and Spindle (Excursion 10), and in these volcanic bombs and cross-bedded base surge deposits have been identified. There is no doubt that originally subaerially exposed tuffs, even with fossil wood fragments, have been carried progressively hundreds of metres down the pipes of many of the vents in what must have been long series of pulses of eruption, collapse and quiescence.

Intrusion and possibly occasional extrusion of basaltic and related magma produced later plugs, dykes and perhaps flows within the vents. Some of these are packed with xenoliths (e.g. Excursion 10) and some extend outwith the vents for some distance, e.g. the Chapel Ness Sill adjacent to the Craigforth Neck at Elie (Francis and Walker 1987, p321).

Vent activity ended at any stage in the sequence and could be renewed many times with different centres of activity occupied in succession, e.g. at Kincraig (Excursion 15).

The general sequence of igneous activity appears to have been as follows:

1. Vent eruption of lavas and tuffs from Strathclyde Group to Coal Measures.

2. Alkali-dolerite sills, some cut by late vents, Limestone Coal Formation to Passage Formation.

3. Main folding and faulting post-Coal Measures.

4. Quartz-dolerite sills and dykes.

5. Early Permian volcanic vents.

References

ARMSTRONG, M., PATERSON, I. B. and BROWNE, M. A. E., 1985. Geology of the Perth and Dundee district. *Mem. Br. Geol. Surv.* Sheets 48W, 48E, 49.

BELT, E. S., 1975. Scottish Carboniferous cyclothem patterns and their paleoenvironmental significance. In Broussard, M.L.S. (ed.) *Deltas, models for exploration.* Houston Geol. Soc., Texas, 427–49.

BINNEY, E. W. and KIRKBY, J. W., 1882. On the upper beds of the Fifeshire Coal Measures. *Quart. Jour. Geol. Soc. Lond.* **38**, 245–56.

BRAND, P. J., ARMSTRONG, M. and WILSON, R. B., 1980. The Carboniferous strata at the Westfield opencast site, Fife, Scotland. *Rep. Inst. Geol. Sci.* 79/11.

BROWNE, M. A. E., 1980a. Stratigraphy of the Lower Calciferous Sandstone Measures in Fife. *Scot. Jour. Geol.,* **16**, 321–8.

———, 1980b., The Upper Devonian and Lower Carboniferous (Dinantian) of the Firth of Tay, Scotland. *Rep. Inst. Geol. Sci.* 80/9.

———, 1986. The Classification of the Lower Carboniferous in Fife and Lothian. *Scot. Jour. Geol.,* **22**, 422–5.

EWING, C. J. C. and FRANCIS, E. H., 1960. No. 3 off-shore boring in the Firth of Forth (1956–57). *Bull. Geol. Surv. Gt. Br.,* **16**, 48–68.

FIELDING, C. R., AL-RUBAII, M. and WALTON, E. K., 1988. Deltaic sedimentation in an unstable tectonic environment – the Lower Limestone Group (Lower Carboniferous) of East Fife, Scotland. *Geol. Mag.,* **125**, 241–55.

FORSYTH, I. H. and CHISHOLM, J. I., 1977. The geology of East Fife. *Mem. Geol. Surv. U.K.*

——— and RUNDLE, C. C., 1978. The age of the volcanic and hypabyssal rocks of East Fife. *Bull. Geol. Surv. Gt. Br.,* **60**, 23–9.

FRANCIS, E. H., 1961. Economic geology of the Fife coalfields, Area II (2nd Ed.). *Mem. Geol. Surv. Scotland.*

———————, 1982. Magma and sediment – I. Emplacement mechanism of late Carboniferous tholeiite sills in northern Britain. *Jour. Geol. Soc. Lond.*, **139**, 1–20.

———————, 1983. Magma and sediment – II. Problems of interpreting palaeovolcanics buried in the stratigraphic column. *Jour. Geol. Soc. Lond.*, **140**, 165–83.

———————, 1991a. Carboniferous. In Craig, G.Y. (ed.) *Geology of Scotland* (3rd Ed.), Geol. Soc. Lond., 347–92.

———————, 1991b. Carboniferous-Permian igneous rocks. In Craig, GY. (ed.) *Geology of Scotland* (3rd Ed.), Geol. Soc. Lond., 393–420.

——————— and HOPGOOD, A. M., 1970. Volcanism and the Ardross Fault, Fife. *Scot. J. Geol.* **6**, 162–85.

——————— and WALKER, B. H., 1987. Emplacement of alkali-dolerite sills relative to extrusive volcanism and sedimentary basins in the Carboniferous of Fife, Scotland. *Trans. Roy. Soc. Edinb.: Earth Sci.*, **77**, 309–23.

GEORGE, T. N., 1958. Lower Carboniferous palaeogeography of the British Isles. *Proc. Yorks. Geol. Soc.*, **31**, 227–317.

GREENSMITH, J. T., 1965. Calciferous Sandstone Series sedimentation at the eastern end of the Midland Valley of Scotland. *Jour. Sedim. Petrol.*, **35**, 223–42.

KNOX, J., 1954. Economic Geology of the Fife coalfields, Area III. *Mem. Geol. Surv. U.K.*

MACGREGOR, A. G., 1960. Divisions of the Carboniferous on Geological Survey Scottish Maps. *Bull. Geol. Surv. Gt. Br.* **16**, 127–30.

PATERSON, I. B. and HALL, I. H. S., 1986. Lithostratigraphy of the late Devonian and early Carboniferous rocks in the Midland Valley of Scotland. *Rep. Br. Geol. Surv.*, **18**, No.3.

WILSON, R. B., 1974. A Study of the Dinantian marine faunas of south-east Scotland. *Bull. Geol. Surv. Gt. Br.*, **46**, 35–65.

———————, 1989. A study of the Dinantian marine macrofossils of central Scotland. *Trans. Roy. Soc. Edinb.: Earth Sci.*, **80**, 91–126.

Chapter 5

Quaternary

Table VII
Excursions 2, 3, 5 and 6

Though no sedimentary rocks of post-Carboniferous and pre-Pleistocene age occur in this area, Geikie (1902, p21) and George (1960, p93) both believed that a cover of very uncertain thickness of Mesozoic and possibly Tertiary rocks extended over part at least of the Midland Valley of Scotland. However, recent work in the offshore areas on both east and west coasts of Scotland suggest that this is not so except perhaps in the Clyde Estuary. There is little to suggest that either Mesozoic or Tertiary sediments were ever deposited in the Fife and Angus areas although Permian sediments do occur only a few kilometres east of the Angus coast (Tayforth sheet 56N 04W 1/250,000 – BGS 1986). Quaternary sediments are, on the other hand, very abundant and these will now be considered.

The area shows signs in many places of a complex glacial history which has been by no means fully worked out. In common with much of Northern Europe the area underwent a series of glaciations during which it was entirely covered by ice. However, the best evidence for this has been obtained not on land but offshore. The surface waters of the oceans contain large populations of temperature-sensitive planktonic foraminifera, both warm- and cold-water species. On death their tests or shells fall to the ocean floor. Successive layers of ocean floor sediment thus point to periods of both warmer and cooler surface waters in the oceans even far from glaciated

areas. The oxygen isotope composition in the calcium carbonate content of the foraminiferan shells is also temperature
related and confirms the evidence from the actual species
present that there have been very many cold periods during
the Quaternary. The evidence points to major fluctuations in
the water masses in the oceans. More specifically, so far as
Scotland is concerned, southward movement of polar water,
and in particular movement of the polar front, is involved
with attendant climatic deterioration. This in turn is likely to
have affected the atmosphere and through it the precipitation
necessary for rapid ice growth (Bowen 1991, p3).

Despite the almost overwhelming evidence offshore for
many glacial stadials or episodes there is within Scotland very
limited evidence of this on land. Interglacial sediments are
rare and difficult to interpret, e.g. in Buchan in North-East
Scotland (Hall and Connell 1991). It appears that the last major
ice sheet, that of the Late Devensian, removed much of the
evidence of what went before from most of the land area of
Scotland. This last major ice sheet lasted from around 26,000 BP
(= before present, taken as 1950 AD) to *c.* 14,000 BP (Bowen
1991, p8). From 1.8 My – the beginning of the Quaternary –
to around 26,000 BP almost nothing remains in Scotland except
in the north-east (Hall and Connell, 1991).

Of the Late Devensian ice sheet we have both direct and
indirect evidence. Sutherland (1991) has summarized this. Except in the north of Lewis the western limits of the ice sheet
are unknown. To the east, the Wee Bankie Moraine in the
North Sea, locally as little as 25km offshore, marks the eastern
limits.

The ice formation was polycentric and the thickness of the
ice sheet is unlikely to have exceeded 1300m. So far as Fife
and Angus were concerned the Eastern Grampians were an
early established centre while the Western Highlands became
increasingly important later. Trewin *et al.* (1987, p47) suggest
that tills containing Scandinavian erratics in North-East Scotland may have incorporated ice-rafted blocks from pre-Late
Devensian times.

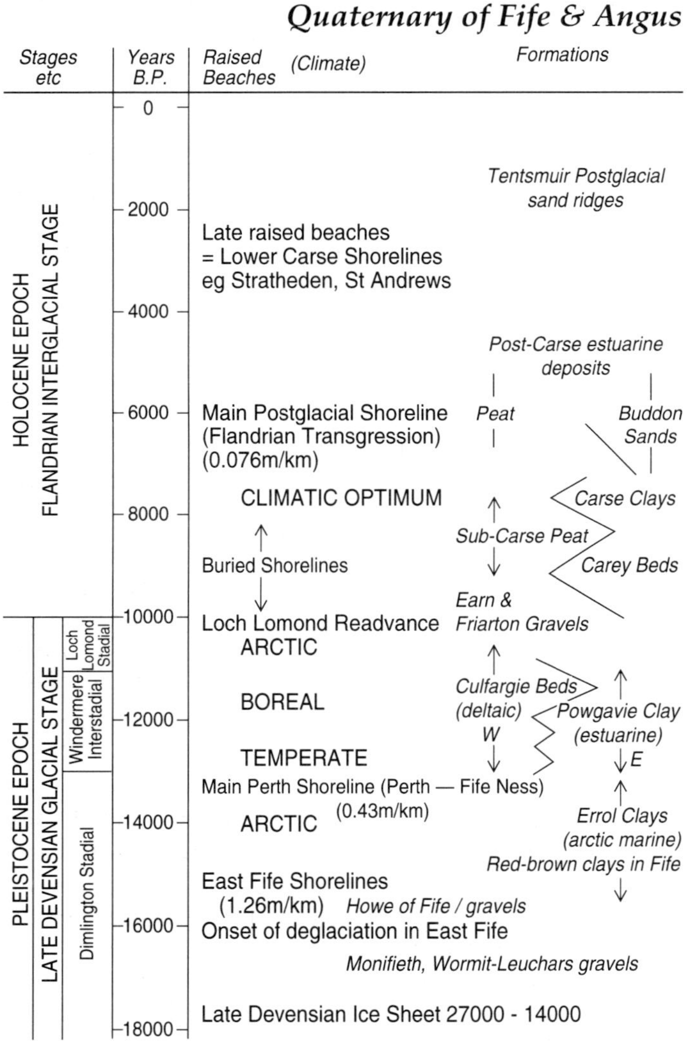
Quaternary of Fife & Angus
Stages etc
Years B.P.
Raised Beaches
(Climate)
Formations
0
2000
4000
6000
8000
10000
12000
14000
16000
18000
HOLOCENE EPOCH
FLANDRIAN INTERGLACIAL STAGE
PLEISTOCENE EPOCH
LATE DEVENSIAN GLACIAL STAGE
Loch Lomond Stadial
Windermere Interstadial
Dimlington Stadial
Tentsmuir Postglacial sand ridges
Late raised beaches
= Lower Carse Shorelines
eg Stratheden, St Andrews
Post-Carse estuarine deposits
Main Postglacial Shoreline
(Flandrian Transgression)
(0.076m/km)
Peat
Buddon Sands
CLIMATIC OPTIMUM
Carse Clays
Sub-Carse Peat
Buried Shorelines
Carey Beds
Earn & Friarton Gravels
Loch Lomond Readvance
ARCTIC
BOREAL
Culfargie Beds (deltaic)
W
Powgavie Clay (estuarine)
E
TEMPERATE
Main Perth Shoreline (Perth — Fife Ness)
(0.43m/km)
ARCTIC
Errol Clays (arctic marine)
Red-brown clays in Fife
East Fife Shorelines
(1.26m/km)
Howe of Fife / gravels
Onset of deglaciation in East Fife
Monifieth, Wormit-Leuchars gravels
Late Devensian Ice Sheet 27000 - 14000

TABLE 7: Late-glacial and Postglacial succession of Fife and Tayside.

Over Fife and Angus, Highland ice is the only ice with which we need be concerned. Glacial striae are found on the more resistant rocks and indicate a broadly eastward flow of ice. The distribution of erratic blocks confirms this, since blocks of Highland metamorphic rocks are found scattered over many parts of Fife and great blocks of quartz-dolerite from the sills further west now lie in many parts of the East Neuk of Fife. In the Carse of Gowrie ice striae are aligned to the ESE gradually swinging round to east by Dundee and slightly north-east by Broughty Ferry. As far north as Montrose there is still an easterly trend for ice movement (Sutherland 1991, Fig. 21, p56).

Glaciers

Within this major ice sheet the following routeways of several distinct ice streams can be recognised.

(i) A Forth glacier lying mainly in the Firth of Forth and flowing eastwards with a branch passing east between the Ochil and Lomond Hills past Kinross and down Stratheden. Drumlinised till in Lower Stratheden and east of St Andrews provide evidence for this (Forsyth and Chisholm 1977, pp233–4). In Stratheden this was augmented by ice from the north side of the North Fife Hills, where striae are from the northwest.

(ii) Carse of Gowrie ice seems to have been derived from Strathearn in the west as well as from the north-west. Blocks of diorite from Glen Lednock are not uncommon in the fluvioglacial deposits at Wormit and indicate a westerly source. Around Perth striae are more commonly from the WNW, as they are too in the Sidlaws. Further east at Dundee, crag and tail features such as Dundee Law point to an eastward movement of ice as do the striae at Wormit. At Broughty Ferry crag and tail features are aligned north of east (Armstrong *et al.* 1985, fig.14).

(iii) Striae in the Sidlaw Hills north-west of Dundee indicate that ice from the Highlands must have crossed Strathmore and the Sidlaws from the north-west. As this flow diminished, ice from the Tay and from much further west must have flowed north-east up lower Glen Isla into Strathmore, a direction confirmed by the north-east-trending drumlins north-west of the Sidlaws (Armstrong *et al.* 1985, p71). Small glaciers in Glens Ardle, Shee, Isla, Prosen, Clova, Lethnot and Esk while active early on (Hall and Connell 1991, p134), latterly do not appear to have made any contribution to Strathmore ice. Red Strathmore till, deriving its characteristic colour from the Lower Old Red Sandstone over which the ice passed, extends for some distance up many of the glens: indeed Bremner (1934) records its occurrence 6.5km up Glen Clova beyond the Highland Boundary Fault and has given similar figures for the Lethnot and Prosen valleys. The limit of penetration has, however, decreased to 1.5km by the time Glen Bervie is reached, thus implying a south-west source for the Strathmore ice.

(iv) To the north lay the large Dee Glacier which occasionally overflowed across the 'watershed' into the northern part of the area normally occupied by Strathmore ice.

Till

Till was defined by Rice (1988, p245) as 'sediment deposited directly from ice' and the term is used here with that meaning.

While ice from different sources can be recognised principally by the nature of erratic blocks that it carried, and by the glacial striae on unweathered bedrock over which it passed, its direction of movement can also be deduced from the orientation of clasts in lodgement till – the till which has been plastered onto the underlying bedrock by the overriding ice. The matrix, often of clay, can also yield information as to its source area. For example, the Strathmore till laid down by ice travelling over the outcrop of the Lower Old Red Sandstone is usually bright red in colour, whereas the till from the ice which travelled over the Carboniferous rocks in Fife contains a great deal of finely-ground, grey shale and even coal and is dark grey in colour. 'Highland' till is often a much paler grey or sometimes yellow-brown. Till usually comprises a range of pebble to cobble sized, usually subangular to sub-rounded, particles in a silty to clay matrix, though this can be sandy where the source is sandstone. The composition reflects the bedrock over which the ice has passed since much of the till is a product of ice erosion, and is often of quite local derivation.

Price has remarked (1983, p66) that 'there are large parts of Scotland where dramatic aspects of glacial erosion are absent'. Over the Fife and Angus area of this guide book this remark is applicable. Rather it is an area of glacial deposition with widespread cover of till or fluvioglacial sand and gravel. Morainic ridges are rare to virtually absent and this is char-acteristic of areas covered by a major ice sheet (Rice 1988, p247) rather than one of valley glaciation. A small number of drumlins have been recognised (Armstrong *et al.* 1985) north and east of Perth, north of Dundee and in Strathearn, as well as those in East Fife. They are aligned parallel to the direction of ice movement.

In East Fife the rather featureless, ice-moulded, drumlinised drift is up to 20m thick but usually much thinner and is often cut through by streams along which the bedrock is now exposed. It is normally brownish-grey in colour, but in North Fife is often more reddish (Forsyth and Chisholm 1977, pp233–5).

Meltwater features

(i) The meltwater from the downwasting Late Devensian ice sheet redistributed much of the till, removing the finest ma-terial ultimately to the sea and reworking the sand and gravel sized material. Material was transported over, through and under the ice in this way and the resulting fluvioglacial

FIGURE 3: Marginal drainage channel, Walton Hill, Cupar, formed by meltwater flowing eastwards at the margin of Stratheden ice during the retreat stages of the Late Devensian ice sheet.

sediments, widespread on both sides of the Tay, are now seen as kettled outwash plains.

In Fife and Kinross such deposits occur around Loch Leven extending almost continuously down Stratheden past Auchtermuchty and Cupar to Leuchars where they merge with those of the Wormit Gap (Excursion 5), associated with Carse of Gowrie ice in the Tay. At Collessie too, fluvioglacial sediments in Stratheden have been augmented by sediment carried southwards by meltwater from the Carse of Gowrie through the Newburgh Gap, past the Loch of Lindores, and spread out round the villages of Collessie and Letham. To the south of the Stratheden sands and gravels there is an extensive spread of sands and gravels from Leslie through Markinch to Kennoway. Fluvioglacial deposits extend up Strathearn to Crieff, lie north of Perth, and are widespread in many parts of Strathmore, e.g. around Coupar Angus, Glamis, Forfar, Brechin, Edzell and Drumlithie where they represent outwash from the Highland glens. On the coast they occur at Monifieth,

Arbroath, the Lunan Water and the Montrose Basin. Eskers, gravel ridges, kame terraces and kettle holes are all well displayed in the Wormit-Leuchars area in north-east Fife (Excursion 6).

(ii) During the melting of the Late Devensian ice, as land appeared above the ice, so many channels were cut in what are clearly now anomalous positions relative to the present-day drainage pattern. Most are now dry or are occupied by small misfit streams. Some were probably ice marginal, others may have been subglacial. Some are very steep and again may have been subglacial. (Sissons *et al.* 1966, p43.) In Fife several occur at Upper Largo, marginal to Forth ice; others are developed at the margin of Stratheden, e.g. at Walton Hill and Cults south-west of Cupar (Knox 1962) (see Fig. 3). A large number were mapped by Armstrong *et al.* (1985 e.g. Fig.14) between Methven, west of Perth, through the Sidlaws, past Dundee to Carnoustie and Arbroath. In the Stonehaven to Aberdeen coastal section several major channels occur apparently coming in from the sea, dropping northwards and returning seawards but still well above sea level (Bremner 1925). Others have been mentioned in the glens tributary to Strathmore by Charlesworth (1955, p784) (Excursion 2).

Raised beaches

Raised beaches are common round the coast (e.g. Excursions 9, 15). They represent old shorelines and indicate that sea level was once relatively higher than it is now. From Table VII, it will be seen that they formed at a number of different times. These old shorelines represent the results of a complicated sequence of events which took place on a number of occasions following the late-Devensian ice maximum. During each period of build-up of ice on land the earth's crust was isostatically depressed by the weight of ice, the depression being greatest where the load of ice was greatest. Simultaneously, abstraction of water from the oceans to form ice sheets, often of vast size such as the Laurentian Ice Sheet in North America and the Scandinavian Ice Sheet in Europe, lowered sea level worldwide by 100m or more. When the ice melted the land began to rise as a result of isostatic readjustment or rebound, but the concurrent rise in sea level was relatively much faster. There was therefore an early rise of relative sea level. Thus early formed shorelines were uplifted as the land continued to rise so that they are often high above present day sea level. Successive shorelines represent usually brief episodes of relative land/sea level stability, long enough for shoreline features to develop. Isostatic readjustment was greatest where the ice was thickest and depression greatest, so that an old shoreline may be expected to have been uplifted more where the ice was thicker. Such shorelines are therefore tilted. Moreover older shorelines will be more tilted than younger ones in the same area, having undergone differential uplift for a longer period of time. This is clearly seen in East Fife where the oldest shorelines have a gradient as they rise westward of 1.26m/km while the gradient of the youngest is only 0.076m/km (Sissons 1974, p328) (Table VII).

The thickest part of the Devensian ice sheet in Scotland lay to the west of the area dealt with in this guide book. Uplift has therefore been greater in the west and individual beaches may be expected to rise when traced westwards. So too within this area younger beaches extended further west than old ones on account of the westward recession of the Late Devensian ice sheet (Browne *et al.* 1981, p8). The arguments for tilted raised beaches were presented by Sissons (1962) and have since been abundantly confirmed by many workers in this area, e.g. Sissons *et al.* (1966) or Armstrong *et al.* (1985). Precise surveying and levelling of beaches at many thousands of sites particularly in the Forth and Tay Estuaries and Strathearn have allowed both widely recognised and quite local raised beaches to be integrated into a late-glacial history of the area, including the stratigraphy of the marine and estuarine sediments of the Tay-Earn area. In addition the micropalaeontology of the sediments and radiocarbon dating of the interbedded peats have allowed detailed palaeoenvironmental and palaeoclimatic analysis of the late Quaternary.

The late-glacial and postglacial history of the area

The oldest raised beaches known in the area are found in East Fife where they extended intermittently from Fife Ness to the St Andrews area. Cullingford and Smith (1966) described these beaches as passing west into fluvioglacial outwash beyond which they cannot be traced. They have also correlated the high Fife Ness to St Andrews shore lines with those of the Anstruther-Leven area but no further west. Cullingford and Smith (1980) presented a further correlation of these East Fife raised beaches with those known between Dundee and Stonehaven, all pointing to westward retreat of the Late Devensian ice sheet and Sutherland (1991) suggested dates of 16,000 to 14,000 BP for the period of ice wasting.

Within the Tay Estuary marine clays frequently lie directly or either bedrock or till and were apparently deposited 'as soon as the westward retreat of the ice-front permitted and continued for an extended period' (Paterson *et al.* 1981, p7). These clays, the Errol Beds, contain a long-known arctic fauna (Paterson *et al.* 1981) and also contain a large number of boulders, many of them striated, and interpreted as dropstones from floating ice. It seems likely (Paterson *et al.* 1981, p7) that the oldest Errol Beds east of Dundee predate the lowest of the old East Fife shore lines of Cullingford and Smith (1966). Errol Bed clays are known as far west as Almond Bridge north-west of Perth suggesting that they were deposited in front of the retreating Late Devensian ice sheet, as a result of a rapid rise in sea level to at least as high as 40m OD near Abernethy. Analogous 'arctic clays', the 'plastic clays' of Forsyth and Chisholm (1977, pp237–47), are present in both the St Andrews and the Elie areas of East Fife. Controversially Browne *et al.* (1981) suggested that foraminifera-bearing, red, laminated clays far to the west at Ladybank in Stratheden also represent deposits contemporary with the Fife Ness shorelines and deposited when the sea invaded the Howe of Fife.

Isostatic uplift resulting from the melting of the Late Devensian ice sheet became more important at this stage (around 13,500 BP) so that relative sea level fell and the ensuing Main Perth Shoreline is widely recognised. It has a gradient of 0.43m/km, slopes down from west to east and has been recognised east from Perth most of the way to Fife Ness and in the Firth of Forth west almost to Stirling. Cullingford and Smith (1980, p28) found that the evidence for this beach north of the Tay 'poses problems'. The sediments associated with this beach, the Powgavie Clay, which is known only from boreholes, are fossiliferous and were deposited under significantly less frigid climatic conditions than the Errol Beds, though still colder than at present (Armstrong *et al.* 1985, p85). Landward the Powgavie Clay passes into the sandy Culfargie Beds exposed around Bridge of Earn and Glencarse-Errol. North of Perth laterally equivalent beds comprise both sand and gravel and are probably fluvioglacial. A number of Lower Perth Shorelines, mainly known from boreholes, point to continuing uplift and thus to falling late-glacial relative sea level. As a result of this fall erosion became widespread, cutting into the late-glacial sediments; indeed channels have been cut to at least 29m below OD, e.g. at Friarton Bridge, Perth. It is believed that this period of erosion immediately preceded the Loch Lomond Stadial (see below) but Armstrong *et al.* (1985, p88) have suggested that gullies cut into the late-glacial sediments might have been formed by headward erosion under periglacial conditions. In this case a Loch Lomond Stadial age would seem likely.

The Loch Lomond Stadial (10,800–10,000 BP), is marked in the west of Scotland by a fresh development of an ice cap and valley glaciers up to 50km in length, the Loch Lomond Readvance. In the East of Scotland pronounced features of this age are lacking but the details have been discussed by Armstrong *et al.* (1985, pp87–9). Suffice it to say that sands (the Carey Beds) and gravels have been assigned to the late stages of this period within the Tay Estuary.

The next widely recognised formation is the diachronous Sub-Carse Peat (9945–7605 BP) with terrestrial vegetation. At Culfargie near Bridge of Earn this peat is at +2m OD and at New Farm, Errol it lies at +4.7m OD, whereas at the Tay Bridges, Dundee it is below sea level. Armstrong *et al.* (1985, p90) suggest that the lowest sea level of the postglacial period was reached at around 8,000 BP. Thereafter eustatic sea level rise was faster than isostatic uplift and caused a widespread marine transgression in east Scotland. This, the Postglacial or Flandrian Transgression, reached its maximum at around 6,000 BP in the Forth, at 6,100 in the Carse of Gowrie, 5,900 in North East Fife, 6,700 at Montrose and 5,700 near Fraserburgh (Price 1983, p161); the resulting Main Postglacial Raised Beach is a widely recognised feature with a gradient of 0.076m/km and extends from Stirling to Fraserburgh (Price 1983, p160). The associated sediments often comprise estuarine clays, silts and sands and are known as the Carse Clays. A basal pebble bed is exposed at the Eden Estuary. The sediments have yielded a largely temperate fauna of invertebrates and have been dated with reasonable accuracy on account of their not only resting on the Sub-Carse Peat but their being overlain at a number of localities by a younger peat. This too has yielded radiocarbon dates, ranging from 6,170 BP (Morrison *et al.* 1981) in the upper Carse of Gowrie to 5,830 BP (Chisholm 1971) at St Michael's Wood, Leuchars. This upper peat over the Carse Clay points to isostatic uplift being once more greater than eustatic sea level rise and hence a renewed fall in sea level. It has resulted in the wide Carselands of the Tay and Forth Valleys with lesser areas for example at the Montrose Basin.

In two areas it has been shown that the Carse Clays give way eastwards to marine sands. This takes place both in the Tentsmuir-Leuchars area (Chisholm 1971) and across the Tay around Buddon Ness (Armstrong *et al.* 1985, p92). Isostatic uplift has continued since the Flandrian Transgression and the formation of the Main Postglacial Raised Beach and lower shorelines have been documented for example in the Tay Estuary (Armstrong *et al.* 1985, fig.15). At St Andrews one such is particularly well displayed on the North Haugh at 3–4m OD. Subsequently uplift seems to have come to a standstill and the tidal gauges at Dundee indicate that sea level has remained constant over the last two hundred years.

References

ARMSTRONG, M., PATERSON, I. B. and BROWNE, M. A. E., 1985. Geology of the Perth and Dundee district. *Mem. Br. Geol. Surv.*, Sheets 48W, 48E, 49.

BOULTON, G. S., PEACOCK, J. D. and SUTHERLAND, D. G., 1991. Quaternary. In CRAIG, G. Y. (Ed.) *Geology of Scotland.* 503–43. Geological Society, London.

BOWEN, D. Q., 1991. Time and space in the glacial sediment systems of the British Isles. In EHLERS, J. *et al.* (Eds) *Glacial deposits in Great Britain and Ireland,* 1–12. Balkema, Rotterdam.

BREMNER, A., 1925. The glacial geology of the Stonehaven district. *Trans. Edinb. Geol. Soc.*, **11**, 25–41.

——————, 1934. Meltwater Drainage Channels and other Glacial Phenomena of the Highland Border Belt from Cortachy to the Bervie Water. *Trans. Edinb. Geol. Soc.*, **13**, 174–5.

BROWNE, M. A. E., ARMSTRONG, M., PATERSON, I. B. and AITKEN, A. M., 1981. New evidence for Late-Devensian marine limits in East-Central Scotland. *Quat. Newsletter*, **34**, 8–15.

CAMERON, I. B. and STEPHENSON, D., 1985. The Midland Valley of Scotland. *Brit. Reg. Geol.* (3rd Ed.)

CHARLESWORTH, J. K., 1955. The late-glacial history of the Highlands and Islands of Scotland. *Trans. Roy. Soc. Edinb.*, **62**, 769–928.

CHISHOLM, J. I., 1971. The stratigraphy of the post-glacial marine transgression in N.E. Fife. *Bull. Geol. Surv. Gr. Br.*, **37**, 91–107.

CULLINGFORD, R. A. and SMITH, D. E., 1966. Late-glacial shorelines in Eastern Fife. *Trans. Inst. Br. Geogr.*, **39**, 31–51.

CULLINGFORD, R. A. and SMITH, D. E., 1980. Late Devensian raised shorelines in Angus and Kincardineshire, Scotland. *Boreas*, **9**, 21–38.

FORSYTH, I. H. and CHISHOLM, J. I., 1977. The geology of East Fife. *Mem. Geol. Surv. Gr. Brit.*

GEIKIE, A., 1902. The geology of Eastern Fife. *Mem. Geol. Surv. Scotld.*

GEORGE, T. N., 1960. The stratigraphical evolution of the Midland Valley of Scotland. *Trans. Geol. Soc. Glasg.*, **24**, 32–107.

HALL, A. M. and CONNELL, E. R., 1991. The glacial deposits of Buchan, north-east Scotland. In EHLERS J. *et al.* (Eds). *Glacial deposits in Great Britain and Ireland*, 129–36. Balkema, Rotterdam.

KNOX, J., 1962. Strand lines at 140 and 190 feet in the Howe of Fife. *Trans. Edinb. Geol. Soc.*, **19**, 120–32.

MORRISON, J., SMITH, D. E., CULLINGFORD, R. A. and JONES, R. L., 1981. The culmination of the Main Post-glacial Transgression in the Firth of Tay area, Scotland. *Proc. Geol. Assoc.*, **92**, 197–209.

PATERSON, I. B., ARMSTRONG, M. and BROWNE, M. A. E., 1981. Quaternary estuarine deposits in the Tay-Earn area, Scotland. *Rep. Inst. Geol. Sci.*, No.81/7.

PRICE, R. J., 1983. *Scotland's environment during the last 30,000 years*. Scottish Academic Press, Edinburgh.

RICE, R. J., 1988. *Fundamentals of geomorphology* (2nd Ed). Longman, Harlow.

SISSONS, J. B., 1962. A reinterpretation of the literature on Late-glacial shorelines in Scotland with particular reference to the Forth area. *Trans. Edinb. Geol. Soc.*, **19**, 83–99.

———, 1974. The Quaternary in Scotland: a review. *Scot. Jour. Geol.*, **10**, 311–37.

———, SMITH, D. E. and CULLINGFORD, R. A., 1966. Late-glacial and Post-glacial shorelines in south-east Scotland. *Trans. Inst. Br. Geog.*, **39**, 9–18.

SUTHERLAND, D. G., 1991. Late Devensian glacial deposits and glaciation in Scotland and the adjacent offshore region. In EHLERS, J. *et al.* (Eds) *Glacial deposits in Great Britain and Ireland*, 53–9. Balkema, Rotterdam.

TREWIN, N. H., KNELLER, B. C. and GILLEN, C., 1987. *Excursion guide to the geology of the Aberdeen area*. Scottish Academic Press, Edinburgh

Descriptive Itineraries

The eighteen excursions in the guide are designed to cover the salient features of the geology of the area extending north from Kinghorn to Stonehaven. They are numbered from north to south and are intended as whole day and half day excursions, the approximate timing being indicated with each itinerary. Each excursion is illustrated by one or more maps on which the localities to be visited are indicated by numbers. There is a brief description in the text for each locality and each itinerary starts with an indication of the walking distance, purpose of the excursion and the route from St Andrews to the area to be examined.

The section on the geology of the area indicates which excursions serve to illustrate the different aspects of the geology. Metamorphic rocks can be seen on Excursions 2 and 4, and the Highland Boundary Fault together with the Highland Border Complex on Excursions 1 and 2. Among igneous rocks, plutonic rocks can be seen on Excursion 4, sills on Excursions 3, 8, 16, 17 and 18; dykes on Excursions 1, 2, 3, 10, 13, 14 and 15; while lavas are well exposed on Excursions 3, 5, 7 and 18. Volcanic necks are abundant in Fife and are magnificently exposed on the coast; they can be examined to advantage on Excursions 7, 10, 13, 14, 15 and 16.

Sedimentary rocks can be seen on almost all the excursions, those of the Lower Old Red Sandstone on Excursions 1, 2 and 3 in particular. The Upper Old Red Sandstone can be seen on Excursions 1, 8 and 17 mainly; the lowest Carboniferous on Excursions 9, 10, 11, 12, 13 and 17 and the Carboniferous

Lower Limestone Formation on Excursions 12 and 18. Quaternary sediments are well displayed on Excursions 3 and 6.

A number of itineraries lead over high ground and these provide good views of the regional geology extending over the north-east part of the Midland Valley of Scotland. Such views are found on Excursions 3, 7, 16 and 17.

Two excursions are particularly suitable as introductory ones. Excursion 9 introduces common sedimentary rock types plus folding and faulting, while Excursion 10 displays particularly straightforwardly volcanic necks and their stages of development.

Which excursions anyone will choose will depend on their interests and time available together with the state of the tide since many excursions are inter-tidal, but the remarks of Sir Archibald Geikie, then Director of the Geological Survey, remain apposite: 'If I were asked to select a region in the British Isles where geology could best be practically taught by constant appeals to evidence in the field, I would with little hesitation recommend the East of Fife as peculiarly adapted for such a purpose. Every teacher of the science appreciates the value of a shore-section where the rocks have been dissected and washed clean and bare by the tides. Round its long stretches of coast-line, the East of Fife presents an almost continuous succession of such sections which for variety, instructiveness, and accessibility have hardly any rivals in the country.' (Geikie, 1902, p iv).

Reference

GEIKIE, A., 1902. The geology of Eastern Fife. *Mem. Geol. Surv. Scotland.*

Crawton Coast Section

Old Red Sandstone Conglomerates
Crawton Basalts
Minor Faults
Cliffs
Dip of strata, angle in degrees
High Water

Buried Pre-Glacial Channel
Car Park
Crawton Farm
Blowhole
Trollochy
Ruin
Track
Ruin
Storm Beach
High Cliff
FOURTH LAVA FLOW
Crawton Bay
THIRD LAVA FLOW
FIRST LAVA FLOW
SECOND LAVA FLOW
0 100 m
0 100 yds

Coast Section East of Arbroath

Raised beach deposits
Upper Old Red Sandstone
Lower Old Red Sandstone
Faults
Cliffs
Dip of strata, angle in degrees
High Water

Seaton Den
Seaton Bay
Deil's Head
Dickmont's Den
Farm road
Arbroath
Stream
Promenade
HWM
Whiting Ness
0 500 m
0 500 yds

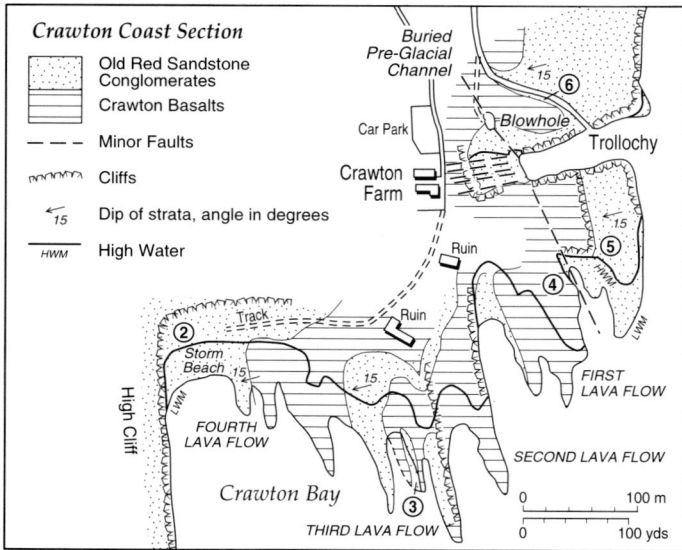

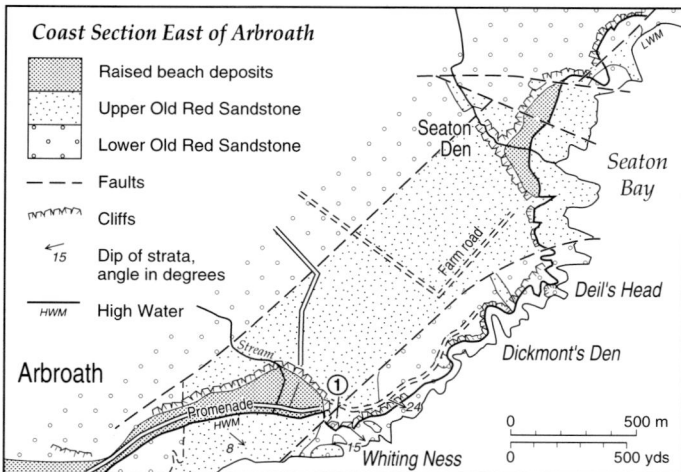

MAP 3: **Arbroath and Crawton**

78

Arbroath, Crawton and Stonehaven (whole day)

OS 1:50,000 Sheets 45, 54
GS One-inch / 1:50,000, Sheets 49, 57, 67
Route maps 3, 4.

WALKING DISTANCE: Arbroath, 1km; Crawton, 1km; Stonehaven 5km; all on rocky shore.

PURPOSE: To examine the following: (1) rocks of the Upper Old Red Sandstone and their unconformable relationship to the Lower Old Red Sandstone; (2) some of the Lower Old Red Sandstone rocks of Kincardineshire, particularly the Crawton Basalts, the Downie Point Conglomerate and the Pridolian (late Silurian) Stonehaven Group rocks at Stonehaven; (3) the Highland Boundary Fault and rocks of the Highland Border Complex at Stonehaven. Structurally the route starts on the south-east side of the Sidlaw Anticline, a NE–SW trending structure of Middle Devonian age. This is gradually crossed and beyond Montrose the rocks form part of the parallel Strathmore Syncline. At Stonehaven the steeply dipping north-west limb of this syncline is met where it abuts against the Highland Boundary Fault. Beyond that lie rocks of the Highland Border Complex and the Dalradian metamorphic rocks of the Scottish Highlands.

ROUTE: Starting from Dundee travel along A92 to Arbroath. This route crosses gently undulating farmland underlain by Lower Old Red Sandstone sediments and lavas belonging to

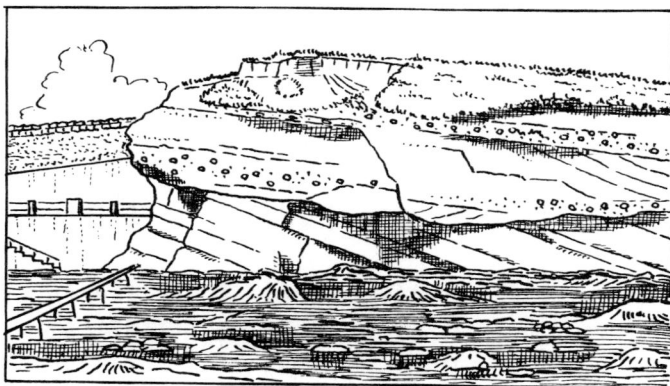

FIGURE 4: Upper Old Red Sandstone conglomerate and sandstone resting unconformably on south-east-dipping Arbroath Sandstone of the Garvock Group, Lower Old Red Sandstone. N end of Arbroath Promenade.

the Arbuthnott and Garvock Groups. On reaching Arbroath pass north-east through the town to the promenade and continue along this to the far end (658411).

1. Arbroath: Whiting Ness; Upper Old Red Sandstone unconformity (Map 3).

In the cliff at the back of the beach red sandstones and conglomerates of the Upper Old Red Sandstone unconformably overlie red sandstones of the Garvock Group Arbroath Sandstone (Armstrong and Paterson 1970, p15). The latter are fine to coarse-grained, pebbly, cross-bedded and with mud-flake conglomerates and belong to the Lower Old Red Sandstone. The line of unconformity can be traced north-east along the cliff for 250m until at the north end the Upper Old Red Sandstone can be seen to be banked against a cliff of Lower Old Red Sandstone. Armstrong *et al.* (1985, p51) have deduced a relief of 100m in the pre-Upper Old Red Sandstone land surface. Within the Upper Old Red Sandstone other sedimentary structures can be seen to advantage. Ramos and Friend (1982, p313) believe these to be alluvial plain deposits. A poorly developed cornstone affecting a conglomerate penetrates the unconformity and has developed a network of carbonate veins in the Arbroath Sandstone beneath. The inclination of the Upper Old Red Sandstone is very gentle (10° ESE) while that of the Lower Old Red Sandstone is 15° to 25° SE (Fig. 4). No fossils have been obtained from the Upper Old Red Sandstone and the beds are assigned to this period on lithological grounds only, resembling as they do the fossiliferous Upper Old Red Sandstone of the Carse of Gowrie (Armstrong *et al.* 1985, p51).

Take the bus north from Arbroath along A92 across very poorly exposed Lower Old Red Sandstone sediments. At Inverkeilor the road rises steeply onto the Ferryden Lavas of the Montrose Volcanic Formation within the Arbuthnott Group. These form higher ground which extends north-east along the strike to the coast just south of Montrose. The low ground of the Montrose Basin is occupied by Quaternary clays and alluvium. These overlie Upper Old Red Sandstone sediments occupying a depression in the unconformity surface over the Lower Old Red Sandstone. Continue north beyond Montrose on A92 until 9km north of Inverbervie and then at 873810 take the unclassified road that runs for 1km south-east to the car park at Crawton Farm on the coast (879798).

2. *Crawton: boulder beach and conglomerate cliff (Map 3).*

At Crawton follow the track that continues south from the end of the road down to the old raised beach cliff. Notice the fine storm beach of boulders derived from the Old Red Sandstone conglomerate. On the west side of Crawton Bay this conglomerate, at the base of the Arbuthnott Group, forms a sheer cliff 30m high with boulders up to 75cm in diameter. Well rounded quartzite is the commonest constituent, but boulders of various granites and lavas, including the Crawton Basalts, also occur, together with greywacke, jasper and 'greenstone' probably from the Highland Border Complex.

In a traverse eastwards along the shore four successive lavas are crossed, the Crawton Basalts. The tops of these flows are highly vesicular and slaggy and the bottoms less so. Sandstone veins penetrate the lavas whose tops are usually weathered, eroded and uneven with the intervening conglomerates lying in hollows on their surfaces. The basalts and the underlying conglomerates belong to the Crawton Group of the Lower Old Red Sandstone and they all dip west into the Strathmore Syncline at about 15° (Armstrong and Paterson 1970, p9).

3. *Crawton: Crawton Basalts and interbedded conglomerates.*

The tops of three flows can be examined on the beach, but only that of the third flow can be seen easily at high tide. The centres of the flows are massive, with columnar jointing. They have resisted erosion and form southward-projecting prom-ontories, with the more easily eroded conglomerates forming narrow bays between them. The following succession can be seen, all thicknesses being very approximate:

	metres
Thick conglomerate (Arbuthnott Group)	—
Fourth Crawton Basalt	12
Conglomerate	6
Third Crawton Basalt	
Thin intermittent red mudstone	15
Second Crawton Basalt	
Conglomerate	3
First Crawton Basalt	9
Thick conglomerate	—

The flows are of porphyritic olivine-basalt with unusually large tabular plagioclase feldspar phenocrysts up to 2.5cm across; these have a strong, flow-banded, parallel orientation thus giving the rock the platy habit characteristic of the Craw-ton Basalts.

4. *Crawton: jointing, amygdales and weathering in basalts.*

At this locality, beside the major NW–SE joint shown on the map, the lowest flow weathers in an unusual manner, the centres of the columns wearing away while the margins stand up to produce a honeycomb rock surface. Campbell (1913, p?40) believed this to be due to silicification along the joints. The lowest flow also contains exceptionally large amygdales up to 15cm across. These contain chalcedony, clear quartz, amethyst and calcite.

5. Crawton: base of the lowest flow and conglomerate.

The base of only the lowest flow is clearly exposed and at this locality, which is reached by crossing the gully at locality 4, it is seen to rest on a fairly flat sandstone surface over the conglomerate, with local irregularities. The conglomerate contains a wide variety of rock types in boulders up to 1.8m in diameter. These include mica-schist, gneisses, schistose grits, quartzite, vein quartz, jasper, pink and grey granites, 'felsite', 'basalt', sandstone and grit. Clast imbrication and cross bedding have been used by Haughton (1988) to demonstrate a southerly and easterly source for much of the Crawton Group conglomerates along the coast. The source area was composed mainly of turbidites cut by calc-alkaline plutonic rocks.

6. Crawton: Trollochy.

On returning to the top of the cliff observe on the shore opposite Crawton Farm a small inlet with vertical sides called Trollochy. A series of closely-spaced E–W vertical faults can be seen at the west end of this inlet and on looking across it from the south side sheared pebbles can be seen on the north face. It is thus apparent that the inlet is due to differential erosion along the faults. The net downthrow is small and to the south – with the eye follow the base of the lava along the cliff. Descend to the head of the inlet where the contact between the base of the First Crawton Basalt and the underlying sediment can be examined. The underlying mudstones are slightly thermally metamorphosed and locally they have been squeezed up into the lava.

A small stream flows over the north edge of the inlet. This is a postglacial stream course, the old course, now filled with

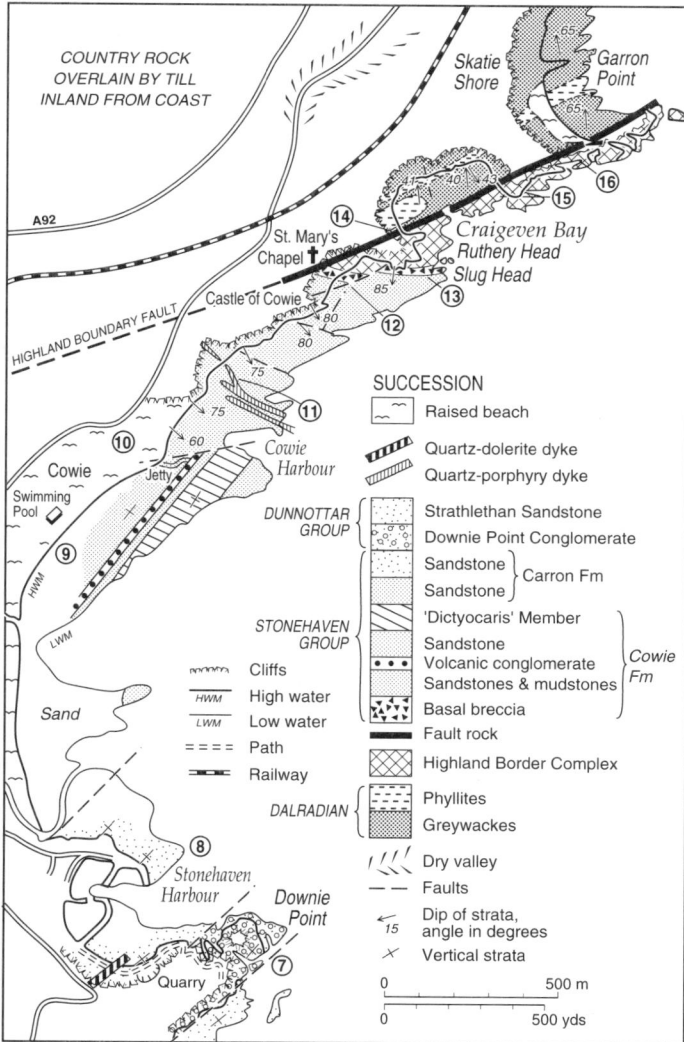

MAP 4: **Stonehaven.**

till, being apparent 60m further west. It lay along the line of the prominent NW–SE joint traversing both the conglomerates and the lavas. A blow hole is occasionally active on it at high spring tides.

Continue northwards along A92 and, for the last 1.5km, A957 to Stonehaven; take the bus down to the harbour and continue eastwards on foot to Downie Point (881851).

7. Stonehaven: Downie Point, conglomerate. (Map 4. The succession for the Stonehaven area will be found on this map.)

The headland is composed of conglomerate at the base of the Dunnottar Group of the Lower Old Red Sandstone (Armstrong and Paterson 1970, p7), the Downie Point Conglomerate. It is cut by large joints and is one of Campbell's (1913) 'highland conglomerates', with boulders of meta-quartzite predominating, but accompanied by granites, porphyries, rhyolites and andesites. It is about 170m thick and has a few impersistent grit bands. The dip is vertical and the top faces south-east, i.e. it lies on the north-western limb of the Strathmore syncline. To the south the conglomerate is succeeded by grey tuffaceous sandstones of the Strathlethan Formation, which have been eroded back to form Strathlethan Bay.

Return to the quarry at the foot of the cliff at Downie Point and examine the base of the conglomerate which can be seen to be remarkably sharply defined. The conglomerate has in fact eroded the sandstones and tuffaceous sandstones of the underlying Carron Formation at the top of the Stonehaven Group. Associated with the tuffaceous sandstones are bands of 'volcanic conglomerate' (Campbell, 1913) in which the pebbles consist predominantly of volcanic rocks, here various acid lavas. In the next quarry to the west cross-bedded red sandstones, at one time worked for building stone, are cut by a 12m thick late-Carboniferous quartz-dolerite dyke running along the strike. This is cut by calcite veins and is deeply weathered at the margins. It can also be examined on the beach.

8. Stonehaven Harbour: Carron Formation sandstones.

If time is available sandstones of the Carron Formation can be examined on the north side of the harbour. These vertical, often massive, brown-weathering, cross-bedded, micaceous sandstones with thin quartzite-pebble beds strike NE–SW and extend along the shore to the mouth of the Carron Water.

9. Cowie Shore; car park.

Take the bus through Stonehaven to the swimming pool (877865) and parking place beside the shore. Dismount and send the bus on to the lay-by (886880) 2km N of Stonehaven.

10. Cowie Harbour; Cowie Formation (730m).

Walk along the shore to the old jetty at Cowie Harbour and follow this out to its seaward end and beyond. There at moderately low tide the highest beds of the Cowie Formation are exposed, namely tuffs and tuffaceous sandstones that overlie the Dictyocaris Member. The latter comprises pale grey micaceous siltstones in a series of red, green and purple ripple-marked sandstones. The relatively rare *Dictyocaris* is to be found in the siltstones and the Cowie Harbour Fish Band lies near the base. The rare fauna, including *Hughmilleria, Kampecaris, Pterygotus, Hemiteleaspis, Pterolepis?* and *Traquairaspis,* indicates a Downtonian or Pridoli (late Silurian) age (Friend and Williams 1978, p19). Throughout this shore section the beds dip steeply south-east, or are vertical, and strike northeast. A few metres from the end of the jetty a prominent trench about 10m wide running along the strike marks the outcrop of a rather soft tuff band cut by a number of small faults. These can best be examined by following a 5cm band of bright green, chloritic sandstone 2m below the base of the tuff band. This shows repeated displacements of 1–2m over a considerable distance along the strike. Just below this sandstone lies a volcanic conglomerate 9m thick and containing pebbles of andesite and more acid lavas. This conglomerate is underlain by a series of brown-weathering sandstones which continue to the top of the beach. Cowie Harbour to the north-east of the jetty lies on the line of a major tear fault across which correlation has not been established.

11. Cowie Shore: sedimentary structures and quartz-porphyry dyke.

For the next 450m along HWM the rocks consist of cross-bedded, brown-weathering, micaceous, grey sandstones showing a number of sedimentary structures which include slumping, mud flakes up to 30cm across and concretions. Two hundred and seventy-five metres from Cowie Harbour a large NW–SE quartz-porphyry dyke complex crosses the shore. It forms a low ridge especially near LWM and can also be traced in the cliff behind HWM at a wooden lookout pole. The dyke is irregular, has several branches and has a reddish, pitted, weathered surface. The fresh rock is pink with creamy feldspars, small micas and quartz crystals, all visible to the naked eye. The dyke has been emplaced along the line of a fault and has baked the surrounding sediment. On the south-west side of the dyke there is a swing in the strike of the beds as they approach the fault, which may imply a dextral displacement.

For the next 350m to the point at Castle of Cowie, the rocks of the Cowie Formation comprise an alternating series of limonitic weathering and tuffaceous sandstones on the one hand and red mudstones and silty beds on the other, the individual units being generally 1.5–3.0m thick. The sandstones form ridges and show slumping, ripples and cross bedding and can occasionally be seen to erode the mudstones beneath. The latter are best exposed in the cliff and show sand-filled mud cracks. At the point one mudstone is completely cut out by the overlying sandstone which contains numerous mud flakes. These sandstones and mudstones comprise fining-upward units and form part of a fluvial complex (Friend and Williams 1978, p19).

12. St Mary's Chapel; basal Old Red Sandstone (Cowie Formation) and unconformity.

In the bay beneath St Mary's Chapel there is much evidence

of minor sinistral and dextral tear faulting, most clearly seen where the faults cross the alternating sandy and muddy beds. The beds strike almost due east-west and the faults strike at about 30° and 110°. For 150m beyond the point exposures are poor at HWM but thereafter it is noticeable that the higher part of the beach has been levelled more than the lower part of the beach. This is because the volcanics of the Highland Border Complex which form the higher part of the beach are more susceptible to erosion than the Old Red Sandstone rocks lower down. The latter can be seen low on the shore but are better exposed in the cliffs of Slug Head to the east where both they and the underlying volcanics are stained a deep hematitic red colour. The fault rock of the Highland Boundary Fault can be seen high on the cliff on the north side of the bay. It is buff-yellow in colour and thus contrasts strongly with the colour of the volcanics.

13. Slug Head: basal Old Red Sandstone and Highland Border Complex volcanics.

The fault running north-east across the bay beneath St Mary's Chapel cuts the cliff 30m west of the first point of Slug Head. To the east the basal beds of the Old Red Sandstone Cowie Formation can be seen both in the cliff and on the shore. They include breccias, containing lava fragments from the Highland Border Complex up to 5cm across, interbedded with siltstones and sandstones. The breccias become less numerous upward in the succession and disappear altogether 75m above the base. Minor faulting is common and the rocks have been considerably sheared, so much so that the lavas in particular are difficult to identify. When fresh lava can be obtained it is seen to be fine grained, greenish-grey in colour and rather crumbly. Henderson and Robertson (1982, p435) stated that the lava chemistry is that of spilitic basalt, i.e. enriched in Na_2O.

At Slug Head the unconformity between the Cowie Formation at the base of the Old Red Sandstone and the Highland Border Complex can be seen, but it is difficult to follow in detail. The Old Red Sandstone rocks form a series of pinnacles and crags striking out to sea and dipping south at about 70°. The contact between them and the underlying volcanics lies at the base of the pinnacles and in the cliff behind. The volcanics have an apparent dip to the north-west, but this is probably due to shear planes parallel to the plane of the main Highland Boundary Fault which lies about 100m to the north-west. Poorly developed pillows are to be seen in a small bay north of that in which the unconformity occurs. They have been affected by shearing adjacent to the Highland Boundary Fault. Near LWM better pillows up to 1m long are displayed and have clear outlines (Trewin *et al.* 1987, p272). **They should not be hammered**. More commonly the lava is crushed and sheared and has largely been converted to chlorite schist,

FIGURE 5: Garron Point and Craigeven Bay, Stonehaven. The rocks in the foreground and on the point in the distance are largely sheared spilitic lavas belonging to the Highland Border Complex. The pinnacle left of centre is composed of carbonated serpentine 'fault rock' on the line of the Highland Boundary Fault. To the left are slightly metamorphosed greywackes of the Dalradian Southern Highland Group.

hence the term greenstone, often applied to the rock. The lavas are continuously exposed on the shore as far as Ruthery Head. Small wedges of black siliceous shales, red jasper and chert also occur, along with stringers of the fault rock described below.

14. Ruthery Head: Highland Boundary Fault.

On the north side of Ruthery Head the fault rock crops out on the beach among lava blocks and boulders. It is a dolomitic and siliceous carbonate, yellow-buff in colour both in fresh and weathered specimens and has a streaky appearance. It is cut by numerous carbonate veins and is believed to be a highly altered serpentine now lying in the line of the fault.

Exposures of the fault rock, the volcanics and the associated shales are much better on the north-east side of Craigeven Bay, the large mass of fault rock there being clearly visible from the south-west side. It is 12m thick, has a sharp 70° north-west-dipping contact with the Dalradian rocks, but a rather diffuse contact with the rocks of the Highland Border Complex on its south-east side.

15. Craigeven Bay, north-east side: Highland Border Complex.

Walk round Craigeven Bay, crossing over Dalradian grits and phyllites of the Southern Highland Group, to the north-east side and cross the thick fault rock. In the small bay 30m to the south-east the strata dip north-west at about 70°, the cleavage and bedding being coincident. Black shales, some of them pyritous and cherty, can be seen interbedded with the volcanics. Individual beds of shale are up to 60cm thick. Similar beds nearby have yielded inarticulate brachiopods and other fossils indicating an Ordovician age for the shales and associated volcanics (Curry *et al.* 1984).

16. Garron Point: Highland Border Complex rocks.

Further exposures of the volcanics, in a schistose condition and cut by carbonate veins, can be seen by returning to Craigeven Bay and climbing up past an old coastguard hut to Garron Point. Many stringers and veins of carbonate cut the volcanics. To the north-west on the Skatie Shore Dalradian rocks are well exposed. A detailed account of these and the section to the north can be obtained in Trewin *et al.* (1987).

Return to the road by walking 450m north along the Skatie Shore to the Den of Cowie, a prominent dry valley spanned by a railway viaduct. Walk up this valley and continue to the road by a footpath which passes under the most northerly arch of the viaduct. The lay-by is 200m to the south-west towards Stonehaven. The dry valley, occupied by the road, is a glacial overflow channel leading in from the sea at 55m and running out again to the north at 30m OD. According to Bremner (1925, p36) it was cut by melt water from the sea ice during the retreat of the last major Quaternary ice sheet.

Return to St Andrews by driving back down into Stonehaven; from the lay-by there is no access to the A90 to the north. From Stonehaven take the A90 to Dundee.

References

ARMSTRONG, M. and PATERSON, I.B., 1970. The Lower Old Red Sandstone of the Strathmore Region. *Rep. Inst. Geol. Sci. No. 70/12*

ARMSTRONG M., PATERSON, I.B., and BROWNE, M.A.E., 1985. Geology of the Perth and Dundee district. *Mem. Br. Geol. Surv.* Sheets 48W, 48E, 49.

BREMNER, A., 1925. The glacial geology of the Stonehaven district. *Trans. Edinb. Geol. Soc.* **11**, 25–41.

CAMPBELL, R., 1913. The geology of southeastern Kincardineshire. *Trans. Roy. Soc. Edinb.* **48**, 923–60.

CURRY, G.B. *et al.*, 1984. Age, evolution and tectonic history of the Highland Border Complex, Scotland. *Trans. Roy. Soc. Edinb.: Earth Sciences* **75**, 113–33.

FRIEND, P.F. and WILLIAMS B.P.J., 1978. *A field guide to selected cutcrop areas of the Devonian of Scotland, the Welsh Borderland and South Wales.* Palaeont. Assoc. International Symposium on the Devonian system.

HAUGHTON, P.W.D., 1988. A cryptic Caledonian flysch terrane in Scotland. *Jour. Geol. Soc. Lond.* **145**, 685–703.

HENDERSON, W.G. and ROBERTSON, A.H.F., 1982. The Highland Border rocks and their relation to marginal basin development in the Scottish Caledonides. *Jour. Geol. Soc. Lond.* **139**, 433–50.

RAMOS, A. and FRIEND, P.F., 1982. Upper Old Red Sandstone sedimentation near the unconformity at Arbroath. *Scot. Jour. Geol.* **18**, 297–315.

TREWIN, N.H., KNELLER, B.C. and GILLEN, C. (Eds.), 1987. *Excursion guide to the geology of the Aberdeen area.* Scottish Academic Press, Edinburgh.

Edzell and Glen Esk (whole day)

OS 1:50,000 Sheets 44, 45
GS One-inch Sheet 66
Route maps 5, 6.

WALKING DISTANCE: 3km footpath, 1km road and 0.5km moorland.

PURPOSE: To examine the following:

1 Sediments belonging to the highest group of the Lower Old Red Sandstone of Strathmore: in ascending order, the Edzell Mudstones (200m), the Gannochy Formation (1400m) and the Edzell Sandstones (180m), together making up the Strathmore Group. The Lintrathen Porphyry belonging to the Crawton Group is also exposed in the River North Esk, but is in faulted contact with the Edzell Mudstones to the south-east and with the rocks of the Highland Border Complex on the north-west side, though this contact could be an unconformity.

2 Ordovician rocks forming part of the Highland Border Complex. Currently they are referred to the Jasper and Greenrock 'Series', which is most probably of Lower Ordovician age, and the Margie 'Series' which, on the basis of its rare fossil content, is known to be of Upper Ordovician (Caradoc-Ashgill) age. Detailed comparisons have been made between these rocks and a more complete succession at Aberfoyle and the reader is referred to the paper by Curry *et al.* (1984) for a

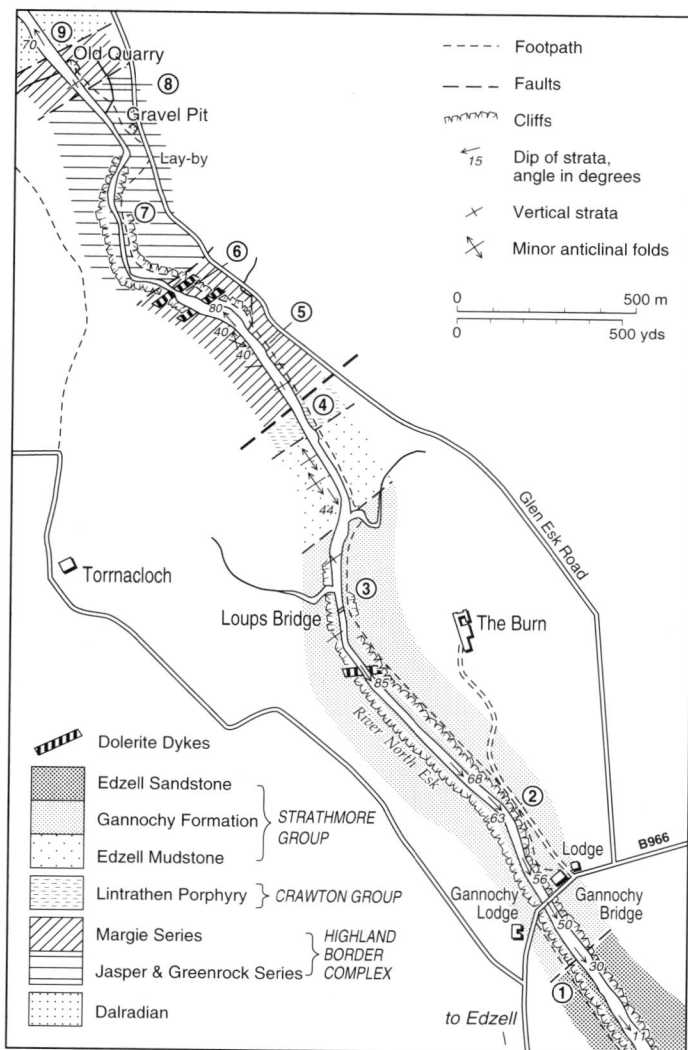

MAP 5: **River North Esk, Edzell.**

full historical review and description. The rocks of the Highland Border Complex are in faulted contact with the Old Red Sandstone to the south-east and with the rocks of the Dalradian to the north-west, and they are themselves cut by several substantial faults, all these faults making up the fault zone of the Highland Boundary Fault.

The Jasper and Greenrock 'Series' comprises spilitic (soda-rich) basalts, 'greenstone' conglomerate, phyllite, jasper and chert, while the Margie 'Series' comprises sandstones, quartzites, shales and the Margie Limestone. It is this last which has, on etching in acetic acid, yielded microfossils which have been used to date the rocks (Burton *et al.* 1984).

3. Rocks belonging to the Dalradian Supergroup. Because of the amount of time involved, following on after the two topics above, the observation of the Dalradian has been deliberately kept to a minimum, one which allows no more than a series of short stops to demonstrate some of the lithologies involved and the rapid rise in grade of metamorphism seen in passing up Glen Esk. Those wishing to examine in more detail the Dalradian succession, Barrow's metamorphic zones, and the phases of deformation which the rocks have undergone, should use the guide to these written by Harte (1987), published in the Geological Society of Aberdeen guide.

Adjacent to the Highland Boundary Fault the Dalradian rocks are greywackes showing graded bedding and occasional cross bedding as the commonest sedimentary structures. These, the Glen Lethnot Grits, belong to the Southern Highland Group and are equated with the Ben Ledi Grits of the Perthshire succession, while the Glen Effock Schists further up Glen Esk are equated by Harte (1979) with the Pitlochry Schists of the Perthshire succession.

Beyond Millden, passing up Glen Esk, the road crosses the Glen Mark Slide and thereafter exposures are in the Tarfside Group, part of the Argyll Group of the Dalradian.

In the Dalradian rocks the grade of metamorphism increases rapidly from the chlorite zone, adjacent to the Highland

Boundary Fault, into the biotite and garnet zones within 1km of the fault. The garnet zone is no more than 1.6km wide as are the succeeding staurolite and kyanite zones, so that high grade metamorphic rocks are rapidly reached in passing up Glen Esk. This is due to the steepness of the disposition of the zones, which Harte *et al.* (1984, p158) have accounted for by D4 deformation in the 'downbend' adjacent to the Highland Boundary Fault.

ROUTE: By bus to Dundee via the Tay Road Bridge, thence by A90 to the Forfar bypass, crossing the Sidlaw Anticline in the Sidlaw Hills, and then travelling to the north and north-east along the Strathmore Syncline by A90 until 3km beyond Brechin. There take B966 north to Edzell. One and a half kilometres north of the village B966 crosses the gorge of the River North Esk at Gannochy Bridge. Dismount at the lay-by beyond the bridge and send the bus on up Glen Esk for 2.4km to a lay-by on the west side of the road (588730).

1. Gannochy Bridge: Strathmore Group sediments.

Take the footpath downstream on the west side of the gorge, noticing conglomerates and sandstones in the Gannochy Formation of the Strathmore Group dipping at 50° south-east in the gorge. Access to these rocks is by means of a gully, 275m downstream from the bridge, where the path bends away from the river for a few metres, or at a flight of wooden steps opposite Gannochy Tower. Hereabouts conglomerates practically die out and this point is taken as the base of the succeeding Edzell Sandstones which consist mainly of rather soft red sandstones and mudstones. The sandstones are ripple marked, contain mud flakes, and are interbedded with blocky red mudstones. A few conglomerate bands are still present but they are thin and the pebbles mainly small. To the southeast the dip decreases until the strata are horizontal some 460m downstream from the bridge.

In the river bank it can be seen that the Old Red Sandstone

is overlain by about 3m of coarse gravels – Quaternary fluvioglacial outwash from Glen Esk.

2. Gannochy Bridge: Gannochy Formation conglomerates.

Return to Gannochy Bridge, cross the river and enter the gateway 20m from the bridge on the north side of the road and follow the footpath upstream on the north-east bank. Much of the gorge is difficult of access and can only be examined in short stretches. It is excavated for the next 1,000m in conglomerates, sandstones and occasional mudstones of the Gannochy Formation. These are all red in colour. Mudstones are well exposed 200m above the bridge in beds up to 2.4m thick and dipping at 55–66°SE. The sandstone bands are often pebbly or gritty, cross bedded and contain large mica flakes. The conglomerates contain a wide variety of rock types. Quartzite is the most abundant pebble type, but porphyry, andesite, granite, felsite and schist are all present. The proportions of the various rock types present vary from one bed to another, and that variation was investigated by Peacock (1961, pp27–33). He found that there was a decrease in maturity, reflected in a decrease in the quartzite + quartz pebble content, upwards through the Gannochy Formation, thus implying great uplift in the source area. Gneiss pebbles constitute up to 25 per cent of the total pebble content in the less mature conglomerates and these rocks also contain considerable amounts of porphyry, granite and sedimentary pebbles, many of which are absent from the more mature rocks. Haughton *et al.* (1990) have argued strongly that Lower Old Red Sandstone conglomerates in south-east Kincardineshire have a northerly source, from an area resembling the Grampian terrane of north-east Scotland, discrepancies being explicable in the unroofing history and the possible involvement of pre-existing conglomerates. Gilchrist (1988) in analysing the conglomerates and their sedimentary structures identified debris flow deposits as the main form present, i.e. those of alluvial fans coming off a mountainous source to the north-west with sufficient sediment being available to overflow

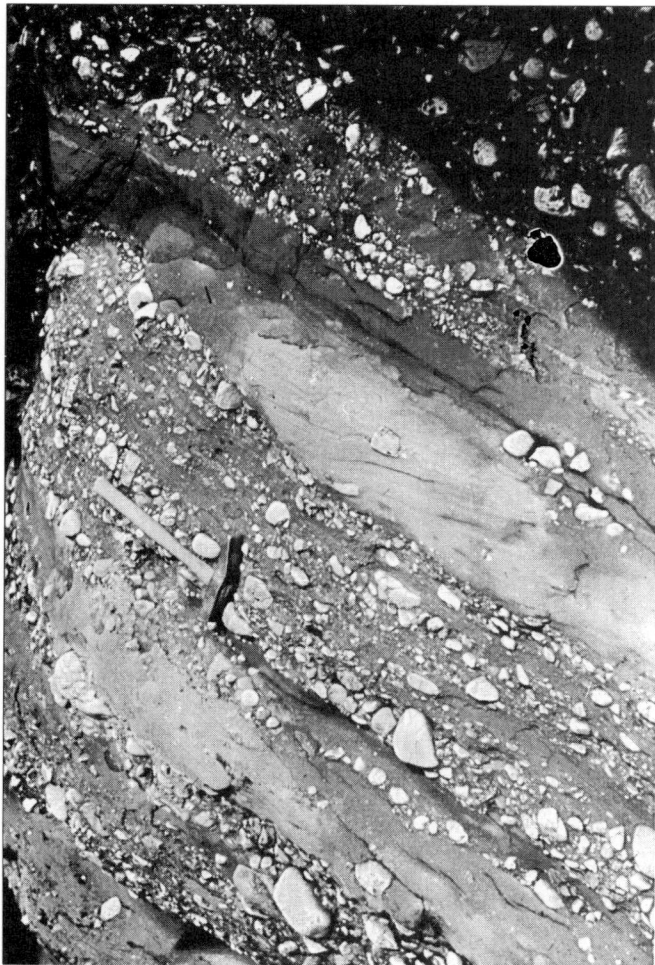

PLATE 1: Conglomerate and cross-bedded sandstone, Gannochy Formation, Strathmore Group, Lower Old Red Sandstone; Loups Bridge, Edzell. Excursion 2, Location 3. (Photo J. A. Weir)

channels on the surface of the alluvial fans. He too postulated a source area comprising principally 'acid plutonic and volcanic complexes intruded into a metamorphic terrane' (Gilchrist 1988, p43).

3. Loups Bridge: Gannochy Formation conglomerate.

The conglomerate, which is now almost vertical, can best be examined at the remains of Loups Bridge, where it is continuously exposed for 230m in the bed and banks of the river (see Plate 1). Thereafter the gorge dies out and the river flows between mounds of fluvioglacial gravels. In this reach of the river the rocks consist of brown limonitic sandstones and mudstones, the Edzell Mudstones, at the base of the Strathmore Group of the Lower Old Red Sandstone, occurring in a series of NE–SW striking minor folds with dips of 30°–75°. Unfortunately exposures, which can only be seen when the river is low, are much better on the west bank which is difficult to reach.

4. Lintrathen Porphyry and Highland Boundary Fault.

The sandstones and mudstones of the Edzell Mudstones Formation continue upstream for 230m to the outcrop of the Lintrathen Porphyry, an ignimbrite at the top of the Crawton Group of the Lower Old Red Sandstone. Commonly weathered to a creamy colour, it is a purplish rock when fresh with small phenocrysts of quartz and biotite visible in the hand specimen. It can be seen at low water 500m upstream from Loups Bridge, abreast of a large erratic block in the middle of the river. The porphyry is believed by Armstrong and Paterson (1970) to be separated from the Edzell Mudstones to the south-east by a major fault and from the rocks of the Highland Border Complex to the north-west by another fault. Neither fault can be seen in the river section.

5. Highland Border Complex sandstones.

Two hundred and seventy-five metres beyond the Lintrathen Porphyry two large Spanish chestnut trees, one on either side of the path, afford a useful landmark. Forty-five metres beyond them large exposures of creamy-grey, gritty, dolomitic and quartzitic sandstones belonging to the Margie 'Series' of the Highland Border Complex can be examined at the water's edge. They are vertical, thin bedded and sometimes micaceous, and a short distance upstream are folded into an anticline plunging to the north-east. Minor puckering is present and small faults cut the sandstone on both banks. (There are two footbridges on the path here within 70m.) For 180m upstream beyond the anticline the dip varies from 20° to vertical, but the general inclination is about 45° NW. Occasional dolomitic, silty or shaly beds occur in the sandstone and there is often a hematitic stain.

6. Dolerite dykes and Highland Border Complex rocks.

At this locality two NE–SW (late Carboniferous) dolerite dykes cut the sandstones, the first about 9m thick and the second about 18m thick. The river narrows here and the second dyke forms bluffs on both banks so that the path climbs steeply at this point. By walking across the dolerite of the second dyke at the river's edge the contact can be seen between the dyke and mid- to dark grey phyllites of the Margie 'Series'. About 15m back from the river and about 10m upstream from the dyke is a crag 5m high of coarse breccia, comprising mainly fine-grained basic igneous rock, but including blocks of limestone with white calcite veins, together with a rather dark grey shaly matrix. Over a century ago this, the Margie Limestone, is said to have been quarried at this locality and also on the west bank. The last 60m up the gorge before it becomes inaccessible at river level comprise purple, limonitic and dolomitic greenstones, strongly crumpled and with steeply plunging minor folds, part of the Jasper and Greenrock 'Series'. Barrow (1901, p330) believed them to be thrust over the sediments.

Return to the path at the top of the gorge and continue upstream. The path after traversing fluvioglacial gravels is cut into greenstones of the Jasper and Greenrock 'Series' for several hurdred metres. Little of the original nature of these rocks can be seen in this vicinity, shearing being intense, but occasional pillow-like forms occur, and Henderson and Robertson (1982, p435) believed them to be pillow lavas. The greenstones are sometimes serpentinous, sometimes phyllitic. There is little to be seen at the riverside hereabouts that cannot be as easily seen along the path except for a 3m lens of dolomitic sandy sediments striking NE–SW exposed at the riverside abreast of an ornamental 'arbour' on the path 280m beyond the thicker dyke.

7. Highland Border Complex; Jasper and Greenrock 'Series'; greenstones, jasper and phyllites.

Fifty-five metres beyond the arbour, where the path runs through a cutting blasted out of the greenstones, there are excellent exposures of red jasper. These commonly comprise isolated masses of jasper up to 1m thick, though others are more persistent and up to 3m across. One at least lies above a fault plane. It seems unlikely that the jasper represents a distinct bed of sediment infolded into the greenstones in iso-clinal folds as was suggested by Barrow (1901). Purple and reddish phyllite bands are present within the greenstones and some of the jasper is associated with these. Black chert is less commonly to be seen. The schistosity strikes north-east and generally dips north-west but is locally vertical, while fault planes have similar attitudes. Beyond the cutting the path swings to the north-east to leave the river and rises until it reaches the lay-by on the road at 588730. The river runs in a small gorge cut in the greenstones for another 600m upstream, the gorge gradually becoming shallower.

8. Highland Border Complex rocks including the Margie Limestone.

From the lay-by walk 140m up the road to a gate on the west side 10m before power wires cross the road. There a track leads past a gravel pit in the fluvioglacial gravels and after 150m crosses a small stream draining swampy ground be-

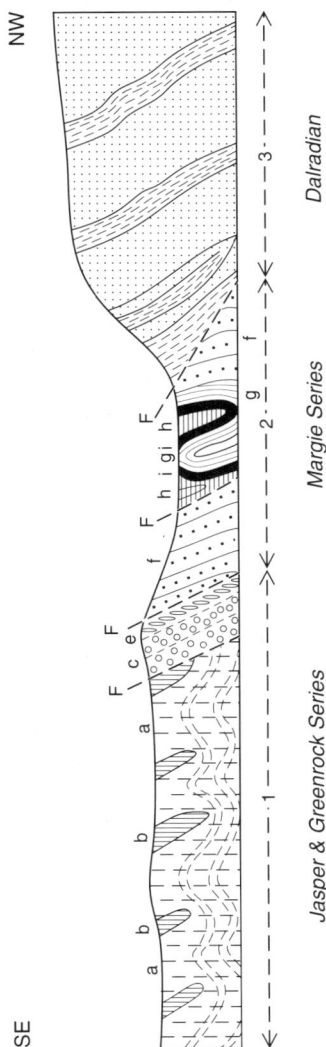

FIGURE 6: Diagrammatic section of some of the rocks in the North Esk (after Pringle, 1942, Fig.1). Length of section = 365m. 1 = Jasper and Greenrock 'Series', 2 = Margie 'Series', 3 = Dalradian Glenlethnot Grits, a = greenstones, b = phyllites, c = greenstone conglomerate, e = sheared gritty rock, f = platy quartzitic sandstone, g = gritty grey shales, h = black shales, i = Margie Limestone, F = faults. The position of the North Esk Fault is also marked.

tween the road and the river. Cross the stream and follow the faint path just above the river bank for 140m upstream until it starts to descend from the river terrace towards a now almost overgrown quarry at the river bank. The exposures for 140m along the bank of the river to just beyond the quarry extend through the rocks of the Highland Border Complex to the North Esk Fault and into the Dalradian greywackes beyond. A similar but by no means identical sequence can be made out on the opposite bank of the river, but it is not advisable to attempt to wade the river even in the dry season.

Working up-river on the north-east bank the sequence consists first of greenstones (a) and then phyllites (b) dipping at 40–60° NW and predominantly green in colour. These are followed by a gap in the exposures which is little more than 30cm wide at low water and believed to be occupied by a fault. Beyond this gap occur 30m of conglomerate (c) with pebbles, up to 12cm in diameter, of greenstone and occasionally jasper in a green matrix. Shackleton (1958, p368) found graded bedding within gritty material in this conglomerate, which indicates younging to the south-east, i.e. that the conglomerate, which dips to the north-west at 70–80°, is overturned. Upstream, 3m of sheared green gritty rock (e) next to the conglomerate marks the upstream limit of the Jasper and Greenrock 'Series' and is followed by about 30m of platy quartzitic sandstone (f) of the Margie 'Series' dipping at 70–85° north-west. Here too Shackleton (1958, p368) found graded bedding younging to the south-east. Thus far there is agreement between the section (Fig. 6) and the exposure. Beyond here there are gaps in the exposures and the interpretation is uncertain. After a gap in the exposures, 8m of creamy, buff-weathering shales are believed to form part of the platy quartzitic sandstone (f) and 2.5m further upstream 5m of buff-weathering grits crop out at the south side of the old quarry.

After a further gap the Margie Limestone (i) is exposed in the river at low water. It is pale grey with quartz grains weathering out on the surface, and cut by frequent white

calcite veins. Almost immediately upstream from the limestone, and also forming a ridge in the quarry floor, are very
dark grey shales and pencil shales, believed to be the black
shales (h) of Pringle's section, where they lie in the small
syncline south-east of the North Esk Fault. The limestone (i)
reappears at the river bank a few metres further upstream to
be followed by gritty, grey shales (g) and quartzose grit with
quartz veins (f) which also forms the north face of the quarry.
Pringle's interpretation of the succession seems to fit reasonably well the exposures seen at low water, but it is by no
means universally accepted: e.g. Shackleton (1958) believed
the succession to be the other way up. The fossil evidence
indicates that the Margie Limestone is Upper Ordovician
while the greenstones are Lower Ordovician in age (Curry
et al. 1984, p124). Units (f) to (i) are all part of the Margie
'Series'.

9. North Esk Fault and the Dalradian rocks.

A gap in exposure beyond the last of the Highland Border
Complex rocks marks the position of the North Esk Fault, the
most north-westerly fault of the Highland Boundary Fault
zone in this section, and is followed by pale grey phyllites of
the Dalradian. The North Esk Fault, Barrow's 'major over-
thrust' (1912, p287), is not exposed on either bank of the river
but separates the Dalradian from the Highland Border Complex. Dalradian rocks are exposed, after a gap, for several
hundred metres up the river and consist initially of semipelites
and then of coarser-grained greywackes, the latter cut by
quartz veins, often of considerable thickness. Dips are consis-
tently to the north-west at 55–85°, but the graded bedding in
the rocks shows younging both up and down the river, thus
implying isoclinal folding, though fold hinges are seldom
seen. These rocks are in the chlorite zone of regional metamorphism.

Return to the bus by the same path.

10. Haughend: Glen Lethnot Grits; garnet-mica-schists.

This locality (573757) lies 100m SSE of the milestone I (Invermark) 11 miles, E (Edzell) 5 miles. The milestone is grey, about half a metre high and is on the west side of the road. There is an area suitable for parking several cars about 50m beyond the milestone and on the east. The nearest bus park is 900m back towards Edzell on the east side of the road (575752).

The exposure consists of several ridges about 1m high of coarse-grained Glen Lethnot Grits within 50m of the road and on the east. Pelitic layers within these are of garnet-mica-schist, the garnets up to about 1mm across. In the grits the earlier fracture cleavage has been refolded on a scale of centimetres to tens of centimetres. The grits carry quartz veins. The dip is 45–75° SE, part of the 'steep belt', south-east of the 'downbend'. The rocks here are already in the staurolite zone of regional metamorphism, but staurolite is scarce or absent at this locality.

11. Mudloch Cottage, Millden: Tarfside Group, sillimanite zone.

From the previous locality drive up Glen Esk through Millden as far as the large lay-by (527779) on the north side of the road where the road has been straightened and opposite the milestone I (Invermark) 7 miles, E (Edzell) 9 miles. This is just 100m short of the track to Mudloch Cottage. Large exposures at the lay-by are dominated by thin-bedded psammites, clean quartzites, usually centimetres thick and often with quartz veins. In the pelitic layers between, muscovite is conspicuous in crystals a few millimetres across, biotite less so. Elsewhere Harte (1987, p206) has pointed out that sillimanite, over much of this ground, is often fibrolitic and difficult to see except by thin sectioning. Structurally the rocks are in the 'flat belt' and are highly deformed. They belong to the Tarfside Group, one in which quartzite is a not uncommon lithology, unlike the overlying Southern Highland Group seen at the previous locality.

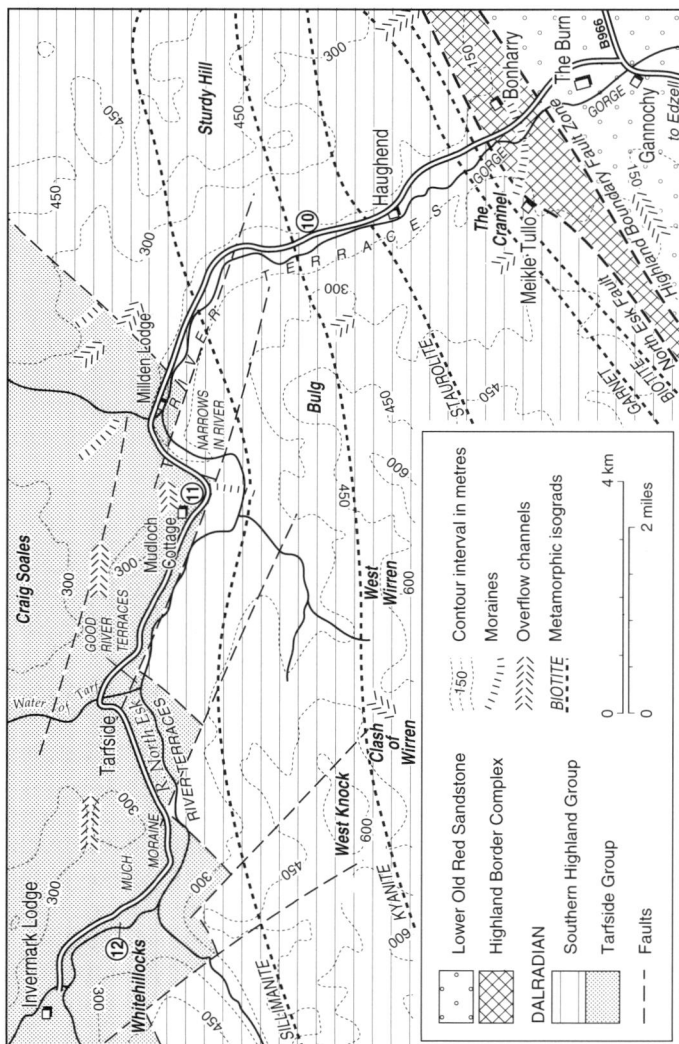

MAP 6: Metamorphic zones, Glen Esk.

Legend:

- Lower Old Red Sandstone
- Highland Border Complex

DALRADIAN
- Southern Highland Group
- Tarfside Group
- Faults

150 ..... Contour interval in metres

,,,,,,, Moraines

>>>>> Overflow channels

BIOTITE Metamorphic isograds

0 — 4 km
0 — 2 miles

12. *Whitehillocks: Tarfside Group, calcsilicate schists.*

Continue up Glen Esk past Tarfside Village, past the entrance to Glen Effock and as far as the footbridge to Whitehillocks Filter Station (453796). There is nowhere to park a bus here, but it can turn 1km further on at the large car park at Invermark (447803) and return by arrangement. Cars can be parked among the trees a short distance beyond the bridge. The footbridge is locked but there are numerous exposures on both sides of the river downstream from it. Quartzites are again present together with pelites and calcsilicate rocks, some cut by calcite veins. These rocks also belong to the Tarfside Group. Locally granitic veins occur and the rocks here too are in the sillimanite zone of regional metamorphism. The foliation dips to the south-east at low angles.

Return to St Andrews by retracing the outward route.

References

ARMSTRONG, M. and PATERSON, I.B., 1970. The Lower Old Red Sandstone of the Strathmore Region. *Rep. Inst. Geol. Sci.* 70/12.

BARROW, G., 1901. On the occurrence of Silurian (?) rocks in Forfarshire and Kincardineshire along the eastern border of the Highlands. *Quart. Jour. Geol. Soc. Lond.* **57**, 328–45.

————, 1912. On the geology of Lower Deeside and the southern Highland Border. *Proc. Geol. Ass. Lond.*, **23**, 274–90.

BURTON, C.J. *et al.*, 1984. Chitinozoa and the age of the Margie Limestone of the North Esk. *Proc. Geol. Soc. Glasgow.*, 124/125, 27–32.

CURRY, G.B. *et al.*, 1984. Age, evolution and tectonic history of the Highland Border Complex, Scotland. *Trans. Roy. Soc. Edinb., Earth Sci.* **75**, 113–33.

GILCHRIST, A., 1988. Sedimentation and provenance of the Gannochy Formation, River North Esk, Edzell. *Unpublished St Andrews Univ. honours thesis.*

HARTE, B., 1979. The Tarfside succession and the structure and stratigraphy of the Eastern Scottish Dalradian rocks. In Harris *et al.*, *The Caledonides of the British Isles — reviewed.* Spec. Publ. Geol. Soc. Lond. **8**, 221–8.

————————, 1987. Glen Esk Dalradian, Barrovian metamorphic zones. In Trewin, N.H., Kneller, B.C., and Gillen, C. *Excursion guide to the geology of the Aberdeen district.* Geological Society of Aberdeen. 193–210, Scottish Academic Press.

————————, *et al.*, 1984. Aspects of the post-depositional evolution of Dalradian and Highlands Border Complex rocks in the Southern Highlands of Scotland. *Trans. Roy. Soc. Edinb., Earth Sci.* **75**, 151–63.

HAUGHTON, P.D.W. *et al.*, 1990. Provenance of Lower Old Red Sandstone conglomerates, SE Kincardineshire: evidence for the timing of Caledonian terrane accretion in central Scotland. *Jour. Geol. Soc. Lond.* **147**, 105–20.

HENDERSON, W.G. and ROBERTSON, A.H.F., 1982. The Highland Border rocks and their relation to marginal basin development in the Scottish Caledonides. *Jour. Geol. Soc. Lond.* **139**, 433–50.

PEACOCK, D.P.S., 1961. The Old Red Sandstone rocks of the Edzell District. *Unpublished St Andrews Univ. honours thesis.*

PRINGLE, J., 1942. On the relationship of the green conglomerate to the Margie Series in the North Esk, near Edzell; and on the probable age of the Margie Limestone. *Trans. Geol. Soc. Glasg.* **20**, 136–40.

SHACKLETON, R.M., 1958. Downward-facing structures of the Highland Border. *Quart. Jour. Geol. Soc. Lond.* **113**, 361–92.

Excursion 3

Dundee to Perth
(whole day)

OS 1:50,000 Sheets 53, 54
GS 1:50,000 Sheets 48W, 48E, 49
Excursion Map 7.

WALKING DISTANCE: Eight kilometres in all with shore walking at Broughty Ferry (1.5km) and hill walking at Mount Quarry (3km) and Dunsinane (1.5km), the rest largely at quarries.

PURPOSE: To examine (1) some of the extrusive and intrusive igneous rocks and sediments of the Lower Old Red Sandstone together with (2) the Upper Old Red Sandstone sediments and finally (3) the late-glacial clays of the area lying between Dundee and Perth. From the hill-top localities a good impression of the structural and regional geology can also be obtained.

ROUTE: By A91 to Guardbridge before following A919 through Leuchars to join A92 and then to the Tay Bridge, 16km. From the bridge follow the signs for A92 to Arbroath for 2.5km east before following A930 for a further 2.5km, signposted for Broughty Ferry, to Douglas Terrace with its narrow road bridge over the railway to the south. Dismount at the promenade (455510) and send the bus back to the roundabout on A930 2km to the west where it should follow Broughty Ferry Road south and east for 500m to a lay-by at which the party will rejoin it (435309).

109

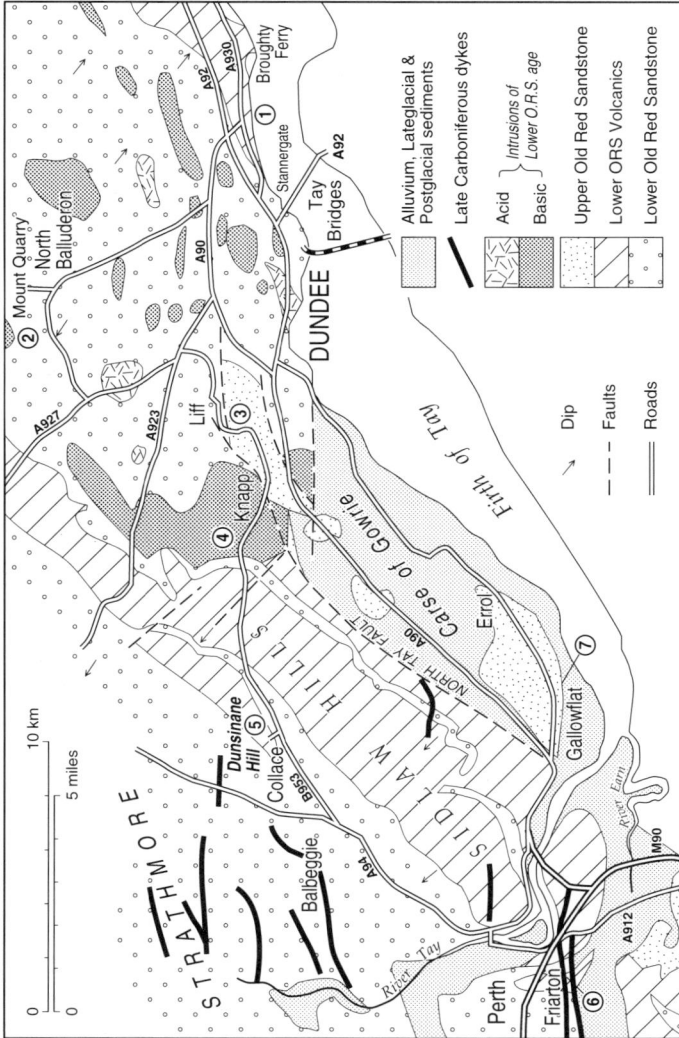

MAP 7: Dundee and Perth district.

Legend:
- Alluvium, Lateglacial & Postglacial sediments
- Late Carboniferous dykes
- Acid } Intrusions of Lower O.R.S. age
- Basic }
- Upper Old Red Sandstone
- Lower ORS Volcanics
- Lower Old Red Sandstone
- Dip
- Faults
- Roads

1. Broughty Ferry – Stannergate shore section: volcanics of the Ochil Volcanic Formation.

This section affords an opportunity of examining both lavas and volcaniclastic sediments within the Arbuthnott Group of the Lower Old Red Sandstone. The Ochil Volcanic Formation is the lateral equivalent of the sediments of the Dundee Formation and indeed the two are interbedded, e.g. at Wormit (Excursion 5) on the other side of the Tay. Structurally the whole section lies on the south-east side of the Sidlaw Anticline and dips are to the south-east throughout.

Walk west for 50m to a flight of steps leading down to the shore. At the foot of the steps are displayed outcrops of south-east-dipping basic pyroxene-andesite lavas, the dip being determined by closely spaced (1cm) joints paralleling the feldspars within the rock. Plagioclase feldspar phenocrysts up to 3mm long occur within a fine-grained mid- to dark-grey matrix which in thin section can be seen to comprise plagioclase feldspar and pyroxene (Armstrong *et al.* 1985, p40).

Return to the promenade and walk 0.5km west to the slipway (448310) at Grassy Beach Outdoor Education Resource Centre. Here a vesicular feldsparphyric basalt flow is exposed immediately west of the slip. It is less closely jointed than the andesite at the previous locality and the feldspar phenocrysts are larger and more abundant. Otherwise there is little difference between the two rocks in the field.

Continue along the beach to the west for 200m to where the promenade is supported on concrete pillars (447309) above outcrops of pyroxene-andesite, much of which is autobrecciated such that blocks up to 2m across lie in an unbedded matrix of angular andesite fragments from sand size up to a few centimetres across. In places the finer-grained matrix is green and chloritic. One hundred metres further west a small headland is composed of unbrecciated andesite though further autobrecciation can be seen in the lower parts of the flow in the remaining 40m of the outcrop to the west. A large brick and concrete beach shelter stands 200m further west along the coast path and 50m beyond this to the west are good exposures of volcanic conglomerate. This comprises both rounded and angular clasts up to 50cm in diameter of vesicular andesite, aphyric andesite and flow-banded andesite in a grey-green chloritic sandy matrix. The bedding is crude and the dip approximately 30° SE. Now walk west again for 400m to rejoin the bus at the lay-by in Broughty Ferry Road.

From Broughty Ferry Road cross the A930 northwards at the roundabout and follow the A972 signposted for Perth. This joins the Kingsway dual carriageway in 700m. Follow the Kingsway west for 4km before leaving it where signposted for Downfield. After 1km bear half left at a roundabout and continue for 5km to the farm road leading to North Balluderon (376387). From here follow the stony and then grassy road north-west on foot for 1.5km, with an ascent of 200m, to reach Mount Quarry, on Cairns Hill in the Sidlaws (363394).

2. Mount Quarry, Sidlaw Hills: sandstones of the Dundee Formation.

Twenty-three metres of strata are exposed in the quarry (Armstrong *et al.* 1985, Fig 6 and pp16–20). Although cross bedding is recorded from the sandstones it is not conspicuous. Plane bedding, often with micaceous partings, is widespread and many of these show current lineation and, in places, evidence of the sandstones sinking into the underlying siltstones; 'ball and pillow' structure resulting from this is widespread and is often well seen in loose blocks on the spoil heaps. Rip-up siltstone clasts are not uncommon in the sandstones. The siltstones display thin sandy partings, starved ripples and ripple cross bedding. Small plant fragments are common and *Parka decipiens* is present.

Armstrong *et al.* (1985) interpreted this type of sequence as in part lacustrine and, where coarsening upward sequences are present, as fluvial, the siltstone clasts being eroded from earlier overbank deposits.

As in all quarries care should be taken when examining the faces and a safety helmet should be worn. Much useful information can be obtained safely from loose blocks.

The south-west end of the quarry is in basic porphyrite, grey-brown in hand specimen, medium grained with scattered vesicles up to 3cm across and occasionally chlorite filled. Under the microscope such rocks usually comprise labradorite feldspar and clinopyroxene. The rock which is columnar jointed is cut off to the south-east by a north-east-trending vertical fault which is well displayed at the south-west end of the quarry. Slickensides on the fault face are mainly horizontal; some are oblique.

On a clear day there is a excellent view to the south from the quarry edge. In the middle distance Dundee Law and Balgay Hill, porphyrite bodies, are conspicuous. To the south-west are the Braes of the Carse comprising, nearest, the Rossie Priory porphyrite sill (Location 4) and, beyond, volcanics of the Ochil Volcanic Formation standing high above the Carse of Gowrie. The Carse is underlain by Upper Old Red Sandstone sediments belonging to the Clashbenny Formation (Location 3), largely covered by the Quaternary Errol Beds (Location 7) and Carse Clays. Beyond the Tay Estuary the Ochil Volcanic Formation, dipping south-east, makes up the North Fife Hills on the far side of the Sidlaw Anticline. Further south still, the peaks of the East and West Lomonds are composed of Carboniferous volcanic plugs (Excursion 16). In the low ground of East Fife the underlying Carboniferous sediments are largely covered by glacial till.

Return by the same route to the bus and drive west through Kirkton of Auchterhouse to join the B954 and travel south to Muirhead. Just beyond the junction with A923 turn south and then west to Liff. In the village turn south for 200m then west for 500m to the Den of Fowlis (328326) and park the bus 100m beyond the narrow bridge at the modern harled white cottages.

3. Den of Fowlis: Upper Old Red Sandstone Clashbenny Formation.

In the Carse of Gowrie the Clashbenny Formation is not seen in contact with the underlying Lower Old Red Sandstone. It is let down between the North and South Tay Faults in a graben within the older Sidlaw Anticline and indeed, although the bright red soils over the Clashbenny Formation are widely known in the Carse of Gowrie, there are few good exposures.

The Clashbenny Formation is exposed in a small stream which joins the main burn in the Den of Fowlis. It is reached from the corner of the field west of the Den and lies just 10m south of the road. Within the wood 30m of the bed of the minor stream consists of bright red, thin-bedded, soft, fine-grained, sometimes gritty, sandstones. They dip at 15° SE and lie a short distance south of an E-W fault within the Tay Graben.

Now take the bus west for 300m, south for 400m and then travel west for 3km to join the road to Knapp and after 1.5km park the bus at the beginning of a track leading to the premises of Knapp Farm Buildings Limited. These stand in a quarry.

4. Hilton of Knapp Quarry (282317): basic porphyrite.

Walk up the road and seek permission to enter at Quarryknowe, the house at the entrance to the quarry. Entry is at one's own risk. The quarry has been excavated in the major Rossie Priory basic porphyrite sill which extends 8km north from Rossie Hill to Adamston Wood. Armstrong *et al.* (1985, p42) reported that chemical analyses of such porphyrites show them to be chemical equivalents of the basic andesite lavas of the Lower Old Red Sandstone. They describe these rocks as of 'doleritic aspect' (ibid. p44).

The rock at the quarry entrance shows good spheroidal weathering, is massive with poorly developed columnar jointing, and grey in colour when fresh. Balsillie (1934, pp135–7) described the rock as a hypersthene-dolerite with the hypersthene inconspicuous in hand specimen. Pink segregation veins, a few tens of centimetres thick and often coarse grained, occur with conspicuous, often albitised, feldspars and with interstitial quartz. Pyroxenes originally present are usually replaced by chlorite.

Return to the bus and drive 3km north-west and west to join B953. Follow this west for 4km before turning north-west for 800m to the large Collace Quarry. (Permission to enter should be obtained from the Roads Department, Tayside Region Offices, Dundee). This part of B953 runs along a WSW-trending strike valley flanked on the north-west side by the steep scarp slopes, in Black Hill and Dunsinane, of the stratigraphically highest part of the Ochil Volcanic Formation, here comprising pyroxene-andesites and overlying sediments of the Arbuthnott Group. Such sediments also bottom the valley while the south-east side of the valley is composed of the dip slopes of pyroxene-andesites also dipping north-west into the Strathmore Syncline.

5. Collace Quarry (208316) and Dunsinane (214317): basalt of the Ochil Volcanic Formation and view of Strathmore.

Collace Quarry is cut into a single 40 metres thick pyroxene-andesite flow, very dark grey in colour and feldsparphyric. Olivine phenocrysts are usually replaced by serpentine and iron oxide, and flow banding parallel to the top of the flow can be seen on weathered surfaces. Fault surfaces within the flow show chlorite development.

Using the footpath that starts from the south side of the quarry ascend to the summit of Dunsinane, one of the vitrified forts of Scotland and one which receives a mention in Shakespeare's *Macbeth*. The summit furnishes an excellent view of the surrounding countryside and its geology. To the east the dip and scarp slopes of the volcanics of the Ochil Volcanic Formation can be seen on Black Hill while to the north-west the ground drops down-dip at 15–20° into the Strathmore Syncline, underlying Strathmore. There the lavas pass beneath younger sediments belonging to the Garvock Group and the topmost group of the Lower Old Red Sandstone, the Strathmore Group. The Syncline is asymmetrical so that when the volcanic rocks of the Arbuthnott Group reappear on the north-west side they are either very steeply dipping to the south-east or are even overturned. Beyond the Highland Boundary Fault, on the north-west side of Strathmore, rocks of the Old Red Sandstone unconformably overlie the Dalradian metamorphic rocks (Excursion 2). In clear weather the higher hills beyond the Highland Boundary Fault, comprising Dalradian metamorphic rocks, can be seen.

Now take the bus back to rejoin B953 and follow this for 5km south-west to Balbeggie and then south-west on A94 for 6km to Perth. There cross the River Tay and follow the A912 south for 1.5km to an unclassified road on the west leading to Friarton Quarry belonging to Wimpey Asphalt, which company should be contacted at Barnton Grove, Edinburgh for permission to enter the quarry (helmets required).

6. Friarton Quarry (116212): basic pyroxene-andesite lavas, inter-bedded sediments and Late-Carboniferous dykes.

From the office follow the quarry road on foot south-west uphill to the higher levels of the quarry in which the lavas are cut by a 35m thick E–W Late-Carboniferous quartz-dolerite dyke. In the lower, northern part of the quarry a 2–3m thick Carboniferous tholeiite dyke forms part of the north face of the quarry and is also conspicuous in the east face. The lavas which occupy most of the north face of the quarry are brecciated adjacent to the tholeiite dyke, the breccia being cemented mainly by pink calcite. On the north side of the dyke in the east end of the quarry the lavas overlie at least 3m of ripple-cross-bedded purplish and greenish siltstones with sandy partings, much of the material being derived from erosion of nearby volcanics. The dip is 15° NW.

In the highest part of the quarry the most conspicuous feature is the 35m thick quartz-dolerite dyke (Armstrong *et al.* 1985, pp57–61). On the south side of the dyke the sediments underlying the lavas are again well exposed in the south-east corner of the quarry. Granules of lava occur in cross-bedded sandstones interbedded with siltstones. Rip-up siltstone clasts occur too. The basal 50cm of the lowest flow has incorporated much sediment, though locally the contact is sharp.

The E–W quartz-dolerite dyke is well exposed in the east side of the quarry in the upper level and displays very coarse, horizontal, columnar jointing which becomes much more closely spaced approaching the margins. In hand specimen the rock is greenish-grey in colour, medium grained with plagioclase feldspar and augite conspicuous on the surface. The chilled margins are noticeably finer grained and darker in colour and are tholeiitic.

Armstrong *et al.* (1985, p57) have summarised the age data for such dykes as 290–295 My or late Westphalian to early Stephanian (= late Carboniferous). Francis (1991, p407), depending on the timescale used, has applied the term Carboniferous-Permian to the suite of intrusions to which these dykes belong. Under the microscope the rocks comprise plagioclase feldspar, augite, pseudomorphs after olivine or hypersthene and a glassy residuum. The tholeiites under the microscope usually have a more glassy matrix while the quartz-dolerites have few or no olivine pseudomorphs (after Armstrong *et al.* 1985, p59). Pink more feldspathic lenses are occasionally exposed in the quartz-dolerite.

From the high ground south-east of the quarry the form of the Sidlaw Anticline can be clearly seen to the south-east. The scarp slopes of NW-dipping lavas and conglomerates form Kinnoul and Kinfauns Hills on the north side of the Tay while 15–25km to the east the scarp slopes of the lavas of the North Fife Hills belong to the south-east limb of the anticline. The intervening low ground of the Carse of Gowrie comprises mainly Upper Old Red Sandstone sediments, let down between the North and South Tay Faults, overlain by Quaternary sediments of the Tay Estuary. Northwards the view across Perth and Strathmore is similar to that from Dunsinane.

From Friarton Quarry drive back to the main A912 road, and return through Perth; cross the River Tay and join A93 to travel southwards before passing the foot of Kinnoul Hill with its spectacular cliffs of volcanic conglomerates at the base and andesite lavas above, all dipping north-west towards Strathmore. At the north end of Friarton Bridge follow A90 east for 7km before following the unclassified road signposted for Errol through Pitfour for 2km to Gallowflat on the south side of the road (212207); the clay pit lies some 500m south-east down the farm road beyond the farm buildings.

7. *Gallowflat Claypit: Errol Beds, Late Devensian Quaternary.*

This locality affords an opportunity of examining the late-glacial marine clays, the Errol Beds, formerly known as the 'Arctic Clays' on account of the polar aspect of the fauna collected in the clays of the same age at Errol 4km further east.

The reddish-brown clays have been subdivided at Errol on the basis of both lithology and faunas, principally of foraminifera, e.g. *Elphidium* spp. and ostracods, a variety of molluscs of arctic aspect and arctic seal. Being further up the estuary the Gallowflat Pit has yielded a sparser fauna. Some 5m of clays and silts are exposed in the pit; bedding can be picked out after weathering and small calcareous nodules are scattered through the clay.

Of considerable interest are the dropstones of various Old Red Sandstone volcanics such as andesites, some autobrecciated, and metamorphic rocks including epidiorites, 'grits' and quartz-veined schists. These dropstones range up to 1.6m across, though most are much smaller, and many are striated. They are interpreted as stones dropped from floating ice, calving from an ice front a few kilometres to the west, and melting in the marine waters of the Tay Estuary. More detailed accounts of the sediments are to be found in Paterson *et al.* (1981), Armstrong *et al.* (1985) and Duck (1990).

To return to St Andrews most directly drive back to the Pitfour-Errol road, cross it and take the minor road leading through Leetown then north-west to A90, back to Dundee and thence via the toll bridge to St Andrews.

Alternatively follow the Errol road through Errol and Kingoodie to Dundee and St Andrews.

Both roads cross the Carse of Gowrie which has few exposures. Most of the Carse is immediately underlain by Quaternary clays (see the chapter on the Quaternary) and below that the Upper Old Red Sandstone Clashbenny Formation, bright red sandstones which form the gentle 'inches' or islands which stood slightly above the peat which at one time covered much of the Carse. The Tay Estuary lies to the south and beyond it are the North Fife Hills of Lower Old Red Sandstone volcanics, e.g. Norman's Law (Excursion 7); they dip south-east and match those dipping north-west in the Sidlaw Hills to the north-west.

References

ARMSTRONG, M., PATERSON, I. B. and BROWNE, M. A. E., 1985. Geology of the Perth and Dundee district. *Mem. Br. Geol. Surv., Sheets* 48W, 48E, 49.

BALSILLIE, D., 1934. Petrography of the intrusive igneous rocks of South Angus. *Trans. Proc. Perthsh. Soc. Nat. Sci.*, **9**, 133–45.

DUCK, R. W., 1990. A study of clastic fabrics preserved in calcareous concretions from the late-Devensian Errol Beds, Tayside. *Scot. Jour. Geol.*, **26**, 33–9.

FRANCIS, E H., 1991. Carboniferous-Permian igneous rocks. In Craig. G. Y. (Ed.). *Geology of Scotland*, pp. 393–420. Geological Society, London.

PATERSON, I. B., ARMSTRONG, M. and BROWNE, M. A. E., 1981. Quaternary estuarine deposits in the Tay-Earn area, Scotland. *Rep. Inst. Geol. Sci.* No. 81/7.

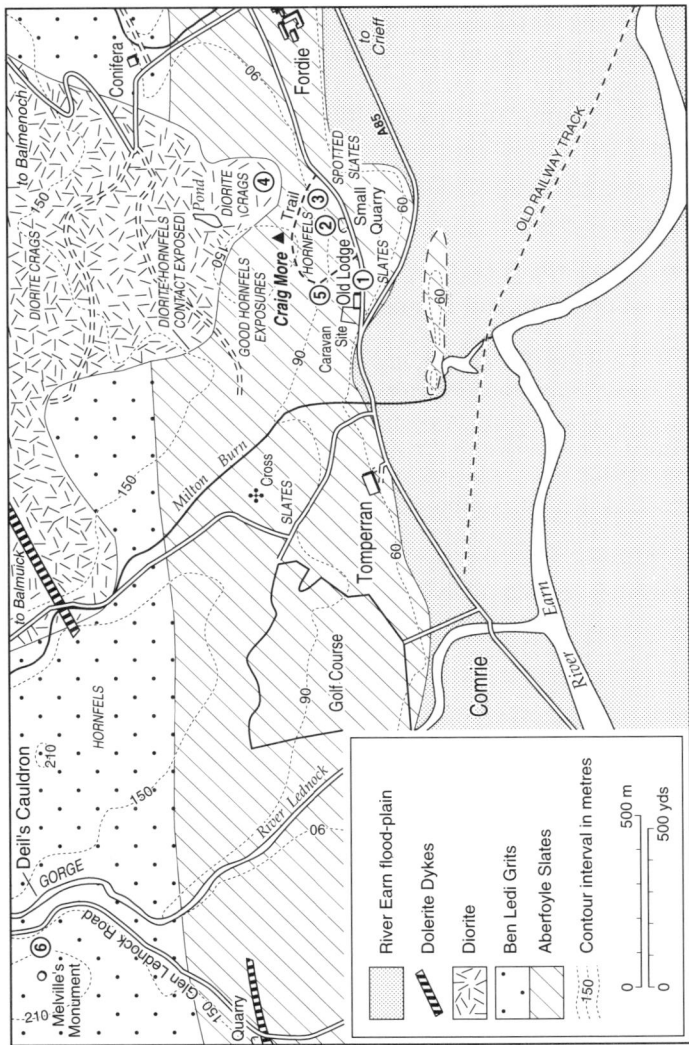

MAP 8: Comrie district.

120

Comrie Igneous Complex (whole day)

OS 1:50,000 Sheet 52
GS One-inch sheet 47
Route Maps 8, 9.

WALKING DISTANCE: Three kilometres of mainly hill walking.

PURPOSE: To examine the Comrie Igneous Complex and its metamorphic aureole in the surrounding Dalradian rocks of the Southern Highlands.

NOTES: The Comrie Igneous Complex (408 ± 5 My, and thus on the Siluro-Devonian boundary) is one of about forty Newer Granites in the Caledonian Province of Scotland. Comrie in many ways is typical of the Newer Granites though in two respects it is unusual. Firstly the pluton is isolated, such plutons commonly occurring in clusters, and secondly its long axis runs counter to the strike of the enclosing Dalradian country rocks.

The Newer Granites are calcalkaline in chemical composition and fit the I–type category of granites. Such granites are typical of the Cordillera and for this reason the Newer Granites have been interpreted as being related to Caledonian subduction, though this view is not universally accepted. The Comrie Complex is part of the South of Scotland suite of granites as it has more affinities chemically with plutons in the Southern Uplands and along strike to the south-west in the Grampians than with the plutons of the Argyll and

Cairngorm suites (Stephens and Halliday, 1984). Many of the Newer Granites show features of compositional zoning, usually from an outer diorite or granodiorite into a core of true granite, and Comrie is again typical in this respect with a dioritic margin and a granite interior. The origin of such granites has been reviewed by Brown (1991).

The major part of the complex is composed of diorites and measures roughly 8km by 3km. It is intruded by a plug measuring 2.5km by 1.5km of fine- to medium-grained pink 'microgranite'. This varies from a pink aplite to a granodiorite composed of quartz, plagioclase, orthoclase, biotite and magnetite. Within the diorite part of the complex two important types can be recognised. (1) One- or two-pyroxene diorite; this is rectangularly jointed, dark grey in colour and contains clinopyroxene, with or without hypersthene, and biotite as the chief mafic minerals. Interstitial micropegmatite occurs and the plagioclase feldspar is sometimes cloudy. (2) A medium- to coarse-grained, grey to pale grey, rough-weathering diorite in which hornblende, probably after pyroxene, and brown biotite are the important dark minerals. Interstitial quartz occurs and locally can be fairly abundant. In addition to these two leading types there are a number of variants including microdiorites containing phenocrysts of andesine feldspar. Their distribution over the complex has not been mapped out in detail.

Within the country rock round the Comrie Igneous Complex two main rock types occur, the psammitic Ben Ledi Grits and the pelitic Aberfoyle Slates, both part of the Upper Dalradian Southern Highland Group, the latter seen only in the south of the area. Structurally they are involved in the 'steep belt', adjacent to the Highland Boundary Fault, and pass north-west into the 'flat belt' (Harte *et al.* 1984). They lie in the nose of the main recumbent Early Caledonoid D1 Tay Nappe, subsequently refolded by the later Caledonoid D4 monoformal 'downbend'. Both these structures trend NE–SW. The steep dips seen in the Aberfoyle Slates and Ben Ledi Grits lie on the south-eastern limb of the D4 monoformal downbend.

To the south-east these highly folded metamorphosed sedi-ments are separated from the unmetamorphosed Lower Old Red Sandstone sediments of the Midland Valley by the High-land Boundary Fault.

The Aberfoyle Slates were said by Tilley (1924, p25) to lie in the chlorite zone of regional metamorphism. Within them, on entering the aureole of the Comrie Igneous Complex, he recognised the following zones of thermal metamorphism: (1) slates (the unaltered country rock); (2) spotted slates in which no chemical reconstruction has taken place, but in which there is a loss of fissility; (3) development of biotite in tiny flakes, which increase in size towards the contact, accom-panied by a further decrease in fissility; (4) a zone of true hornfelses with complete loss of fissility and appearance of cordierite, giving the rock a purplish appearance on a fresh surface. Zone 2 appears about 425m from the contact, Zone 3 at about 270m from the contact and Zone 4 at 135m from the contact. Within Zone 4 Tilley distinguished among others a silica-rich type characterised by cordierite and hypersthene, and a silica-poor type characterised by cordierite, spinel, co-rundum and sometimes biotite. These two types are not dis-tinguishable in the field.

Induration is generally the first sign of thermal metamor-phism in the Ben Ledi Grits, spots appearing concurrently in the interbedded argillaceous bands. Biotite replaces chlorite and white mica in the vicinity of iron ore at about 450m from the contact with the diorite. At 250m from the contact true hornfelses appear, andalusite appearing in the argillaceous bands along with cordierite, the biotite now occurring in large flakes. At the contact the hornfels is coarse grained and almost of granitic aspect with conspicuous biotite. Original sedimen-tary banding may survive close to the contact as may the fracture cleavage seen in the Ben Ledi Grits along almost the whole length of the Highland Boundary Fault.

At the northern end of the diorite the progressive thermal metamorphism of an epidiorite (a regionally metamorphosed dolerite) can be studied. Outside the aureole the epidiorite is composed mainly of hornblende, sodic plagioclase and zoisite, with minor epidote and chlorite. Approaching the diorite, magnetite appears in the hornblende and the small hornblendes change into biotite while zoisite and epidote disappear and the plagioclase becomes more calcic. Finally, close to the contact the rock becomes a granular pyroxene–hornfels containing hornblende, diopsidic pyroxene, hypersthene, labradorite, biotite and iron ores.

ROUTE: Follow A91 westwards to Cupar (16km) then by A913 north-west through the North Fife Hills (Excursion 7) composed of Lower Old Red Sandstone volcanics, to Newburgh (16km) overlooking the Tay and the Carse of Gowrie beyond. The North Fife Hills composed of SE-dipping volcanics are matched on the other side of the Tay by Lower Old Red Sandstone volcanics dipping to the north-west in the Sidlaw Hills on the other side of the NE–SW trending Sidlaw Anticline. After 7km join A912 for 2km leading to M90. Notice the spectacular road cut on Moncreiffe Hill in Lower Old Red Sandstone basic pyroxene-andesites. Follow M90 round the west side of Perth to A9 before leaving on A85 which runs along the north side of Strathearn to Crieff (27km). Note the very red soil derived from the Lower Old Red Sandstone, here in the centre of the Strathmore Syncline. The south side of Strathearn is marked by the Ochil Hills composed of volcanic rocks dipping north-west into Strathmore.

From Crieff travel west on A85 for 8km to a caravan site on the north side of the road (785225) on the outskirts of Comrie, and dismount.

1. Craigmore: Aberfoyle Slates, aureole of the Comrie Igneous Complex

On the main road walk 50m east from the caravan site to a gap in the wall and a wooden gate on the north side of the road. This leads to a forest trail which starts on the old drive leading east to Fordie. One hundred and fifty metres east of the end of the drive examine the Aberfoyle Slates in a small cutting. They are fine grained, light grey in colour and the cleavage dips steeply north. There is a horizontal lineation cn the cleavage surfaces. The slates lie in the chlorite zone of regional metamorphism and there is no trace of thermal metamorphism at this point.

2. *Arenaceous bands in the Aberfoyle Slates; spotted slates*

A small quarry 180m further east on the north side of the road has good exposures within the slates with silty and sandy bands showing original sedimentary structures and quartz veins and blebs in the more siliceous rocks. The bedding dips 80°–85° N and the cleavage 75° N. Spotting, indicating the first stages of thermal metamorphism, is present in the argillaceous rocks, but is poorly developed and not obvious.

Forty-five metres beyond the quarry a 3.5m high bluff of slates on the north side of the road is partly covered by a large *Cotoneaster* bush and here spotting is well developed. On fresh surfaces the spots are slightly darker than the surrounding rock and are up to 1mm in diameter.

3 *Spotted slates*

At a left hand bend in the road 120m east of the quarry slates show larger spots up to 3mm across and the rock is noticeably less fissile than hitherto. Thin silty bands occur and are often crumpled.

4. *Hornfelses and the hornfels-'diorite' contact*

Continue east along the Fordie driveway for a further 150m and then take the forest walk steps which lead north-west to the summit of Craig More. The first exposures, reached after following the path for about 250m to a large fallen tree, form part of a large bluff mainly on the right of the path. They comprise fine-grained, now purplish, splintery hornfels with none of the original fissility of the slates. Thin sandy and silty siliceous bands make up about 30 per cent of the rock and isoclinal folds can be seen in many of the exposures. The purplish colour, common in cordierite-bearing hornfelses, marks a much higher grade of thermal metamorphism and is found within 100m of the contact with the diorite. At the root of the fallen tree is a 1m porphyrite dyke (trend 170°) with the contact well exposed and chilled margins. The pitting on the surface of the dyke corresponds to feldspar phenocrysts within it.

Follow the path for 180m to the summit of Craig More and a bench seat fixed to a large, clean, glaciated exposure of hornfels. From this seat there is an excellent view of Strathearn and of Glen Artney lying on the line of the Highland Boundary Fault with SE-dipping sediments of the Lower Old Red Sandstone Arbuthnott Group beyond.

To reach the contact between the hornfelses and the diorite, leave the path and walk 60m north-east to a large oak tree at the head of a gully in the east-facing crags. Descend this gully for 20m and just to the north will be found the contact. The hornfels, showing strong banding, overlies a pinkish more granodiorite-like development of the Comrie Complex. This is readily recognised by its quite different blocky jointing and pinkish colour. Hornfels xenoliths are common. Near the contact, which is irregular, the hornfels is much coarser grained and can be difficult to separate from the granodiorite. Grey D2 hornblende-diorite is exposed 200m further east in low knolls overlooking a field to the east.

5. Biotite hornfels

Return to the seat and then follow the path west along the summit of Craig More for 250m to where the path turns south and 70m downhill to another bench seat, also on a fine glaciated knoll, this time on medium-grade biotite hornfels, still hard and splintery but no longer with a purplish hue, and in which bedding and quartz veins can be seen.

Now follow the fenced path for 200m down to another glaciated knoll level in height with the tops of the large redwood trees lining the Fordie driveway below. The knoll comprises spotted slates, once again fissile, and much further from the outcrop of the Comrie Igneous Complex. They are cut by a NE–SW trending columnar-jointed lm wide dolerite dyke. Contrast the jointing of the two rock types. Now follow the path back down to the road (l90m) and rejoin the bus.

6. Glen Lednock: Deil's Cauldron, hornfelses

Take the bus through the town of Comrie to follow the Glen Lednock road north-west for 1.5km to a lay-by where the bus can be parked (767235). This is just beyond the footpath leading up to Melville's Monument. The monument, high above the road to the south-west, stands on slates and siltstones, within the Ben Ledi Grits and showing no evidence of thermal metamorphism. Take the signposted footpath and staircase down to Deil's Cauldron overlooking a waterfall in the River Lednock. Although the cauldron is some 500m from the mapped margin of the Comrie Igneous Complex, the Ben Ledi Grits, still with fracture cleavage, are now high grade rather siliceous hornfelses, and the clear implication is that the roof of the complex must slope outwards at gentle angles such that the contact lies at no great distance beneath the surface. The fracture cleavage is no longer a plane of parting, on account of the thermal metamorphism that the rocks have undergone.

7 Shaky Bridge: marginal diorite

Take the bus on up the glen for 500m to where the footpath leading to Shaky Bridge and Balmuick leaves on the right, and walk 230m down the path to the bridge (763243). At this point there are good exposures of the marginal, coarse-grained xenolithic D2 diorite, a rock containing conspicuous biotite and interstitial quartz together with andesine feldspar and amphibole after clinopyroxene.

8. Funtullich: high-grade hornfels, microgranite, microdiorite

Return to the bus and continue up the glen, noticing the rock barrier across the glen, 800m above Shaky Bridge, through which the River Lednock has cut a small gorge. There is a considerable amount of hummocky moraine in this vicinity.

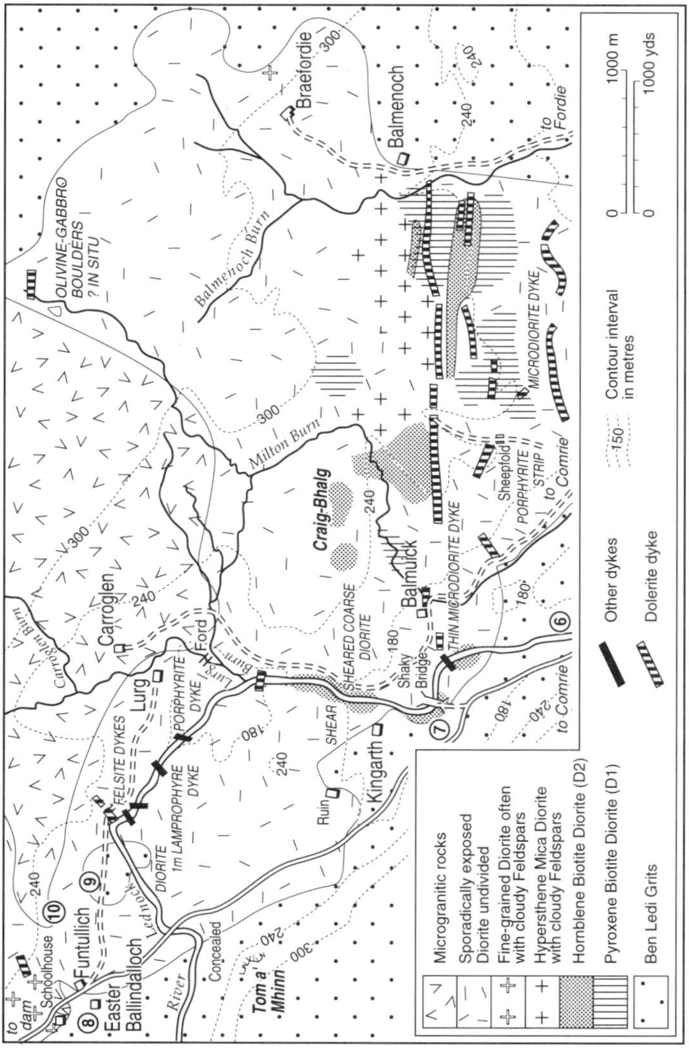

Microgranitic rocks
Sporadically exposed Diorite undivided
Fine-grained Diorite often with cloudy Feldspars
Hypersthene Mica Diorite with cloudy Feldspars
Hornblene Biotite Diorite (D2)
Pyroxene Biotite Diorite (D1)
Ben Ledi Grits
Other dykes
Dolerite dyke
Contour interval in metres
150
0
1000 m
0
1000 yds
to dam
Schoolhouse
Easter Ballindalloch
Funtullich
River
Lednock
Concealed
Tom a' Mhinn
DIORITE
1m LAMPROPHYRE DYKE
FELSITE DYKES
PORPHYRITE DYKE
Lurg
Carroglen
Carroglen Burn
Lurg Burn
Ford
Ruin
Kingarth
SHEAR
SHEARED COARSE DIORITE
Shaky Bridge
Balmuick
THIN MICRODIORITE DYKE
Craig-Bhalg
Milton Burn
Sheepfold
PORPHYRITE STRIP
MICRODIORITE DYKE
to Comrie
OLIVINE-GABBRO BOULDERS ? IN SITU
Balmenoch Burn
Braefordie
Balmenoch
to Fordie

MAP 9: Part of the Comrie Igneous Complex

In the last 800m before reaching Funtullich a broad alluvial plane on either side of the road obscures the west margin of the diorite. At Funtullich 36m beyond the farmyard wall a tongue of high grade hornfels crosses the road and runs on into the diorite for 90m to the north-east. The hornfels is highly quartzose and contains subordinate feldspars and biotite. It is cut by veins of microgranite. This rock also crops out immediately to the north-west of the hornfels and also west of the road and can be distinguished by its widely spaced, flat lying or gently dipping joints. This is an isolated outcrop of microgranite lying some distance from the main plug, and its relationship to the latter is obscure, as are those of several adjacent small patches. The microgranite may be examined on the west side of the road in front of the small, isolated stone house, the old school house. The rock is pinkish in hand specimen and consists principally of quartz, oligoclase feldspar and biotite with some micropegmatite. Small xenoliths of hornfels are common. Behind the small house and 22m to the west finer-grained, dark grey microdiorite crops out beyond the microgranite. It is cut by a 30cm thick pink microgranite vein with sharp contacts but no chilled margins. A similar microdiorite occurs at the roadside level with the house. One hundred and eighty metres north-west of the house, beyond an old gravel pit, the microdiorite is paler in colour, porphyritic, and contains small hornfels xenoliths. In thin section the microdiorite often bears cloudy feldspars with clear rims, biotite and pale green amphibole. Hypersthene occurs near the contact with the country rock, and quartz, zircon and sphene occur as accessory minerals.

9. East of Funtullich: cordierite hornfels and margin of microgranite

Follow the track east from Funtullich for 450m to a gate through the wall on the north side. Forty-five metres north of the gate rock ridges trending NE–SW are composed of high-grade cordierite hornfels cut by veins of microgranite and diorite. The purplish colour is particularly noticeable on fresh surfaces of the hornfels. The rock ridges apparently conform to the strike and original sedimentary banding in the country rock. This may be part of a very large inclusion within the diorite, but lack of exposures precludes proof of this. Continue to the north passing about 30m west of the solitary tree in this field across diorite of variable composition and then 140m north-west to a gate in the wall. Just south of the gate the margin of the main microgranite mass is reached. The rock here is quite fine grained, but on the north side of the wall it is much coarser in grain and very similar in composition and appearance to the microgranite at Funtullich.

10. East of Funtullich: diorite and microgranite contact

Ninety metres to the west and just above the wall is a rocky bluff with a single tree growing in a crack in the rock. At this point the steeply dipping contact between the west margin of the microgranite and the main diorite mass can be seen in a 30cm deep gully in the face of the crag at the base of the tree. The microgranite is again similar to that at Funtullich while the diorite is medium- to coarse-grained, the grain size increasing to the west. Xenoliths of both diorite and hornfels occur in the microgranite and pink almost aplitic veins cut the diorite, the microgranite and some of the xenoliths. Conspicuous coarse joints in the diorite dip gently to the east. Looking up the hillside to the north-east the approximate line of contact between the diorite and the microgranite is indicated by the colour contrast between the dark grey crags of the former and the pale pink of the latter. Return to the bus by continuing along the wall to the west past outcrops of coarse-grained diorite till the corner of the wall is reached and then walking south-west to Funtullich.

11. Glen Lednock Dam: epidiorite and porphyry dykes

Take the bus up the glen for 3km past Invergeldie, noticing the prominent rock face of Spout Rolla in front of the dam. Formerly a waterfall, this is now usually dry. It lies within the diorite aureole and comprises hornfelsed grits within the 'flat belt' (Harte *et al.* 1984). Leave the bus at the start of the large Z-bend just above the bridge over the Lednock (728286). The epidiorite at the bridge is part of a regionally metamorphosed dolerite sheet intruded into the Ben Ledi Grits before the regional metamorphism of the area, i.e. it is one of the Older Basic Intrusions within the Dalradian rocks (Johnston 1966, p45). It is very varied in appearance and occupies much of the bed of the river between the bridge and the dam. A contact with the country rock can be examined on the west bank of the river at a small waterfall 10m downstream from the bridge, but it is apparently sheared. Upstream different types of epidiorite recur in many exposures, some being very coarse grained with amphibole crystals up to 2cm long set in a more feldspathic matrix. These large amphiboles sometimes show a strong preferred orientation. Elsewhere the rock is medium- or fine-grained but contains dark streaks and blotches of the coarser-grained material. McDonagh (1964) believed the epidiorite sheet to consist of a lower coarse-grained member and an upper medium-grained member intruded separately. Pink feldspathic streaks occur in the rocks and they too are traversed by calcite, quartz and amphibole veins.

Several dykes can be seen cutting the epidiorite in the bed of the river. One on the east bank 10m above the bridge shows a margin controlled by joints in the country rock and is chilled against the epidiorite. In some, stringers of porphyry run out into the country rock, e.g. beneath the bridge where the dyke rock is fine grained, greenish in colour and is a porphyry with phenocrysts of pale amphibole and feldspar in a fine micro-crystalline matrix. Porphyry dykes of this type commonly occur in association with the Caledonian Newer Granites (Johnston 1966, p51). Exposures of the epidiorite can also be seen at both ends of the dam, locally with clots of epidote and, again, cut by porphyry dykes with chilled margins.

Where the epidiorite enters the aureole of the Comrie igneous complex it has undergone progressive thermal metamorphism. Thermally metamorphosed epidiorite can be seen

400m east of Spout Rolla waterfall and above the road, but it is not well exposed.

Return by bus down Glen Lednock to Comrie and by the same route back to St Andrews.

References

BROWN, P. E., 1991. Caledonian and earlier magmatism. In Craig, G.Y. (Ed.) *Geology of Scotland,* pp.229–95. Geological Society, London.

HARTE, B. *et al.,* 1984. Aspects of the postdepositional evolution of Dalradian and Highland Border Complex rocks in the Southern Highlands of Scotland. *Trans. Roy. Soc. Edinb.: Earth Sci.,* **75**, 151–63.

JOHNSTONE, G. S., 1966. The Grampian Highlands. *Brit. Reg. Geol.* (3rd Ed.).

McDONAGH, M. J. O., 1964. The metamorphism of a basic sill. *Unpublished St Andrews Univ. honours thesis.*

STEPHENS, W. E. and HALLIDAY, A. N., 1984. Geochemical contrasts between late Caledonoid granite plutons of northern, central and southern Scotland. *Trans. Roy. Soc. Edinb., Earth Sci.* **75**, 259–73.

TILLEY, C. E., 1924. Contact-metamorphism in the Comrie area of the Perthshire Highlands. *Quart. Jour. Geol. Soc. Lond.* **80**, 22–70.

Excursion 5

Wormit Shore (half day)

OS 1:50,000 map, Sheet 59
GS 1:50,000, sheet 48E
Excursion map 10.

WALKING DISTANCE: 3.25km on rocky and shingle shore.

PURPOSE: To examine a sequence of volcanic rocks, together with minor amounts of sediments, belonging to the Ochil Volcanic Formation of the Arbuthnott Group of Lower Old Red Sandstone age, exposed on the southern shore of the Tay Estuary at the Tay Railway Bridge. The igneous rocks consist of a rhyolite body and associated breccia; a number of lava flows, mainly andesites, agglomerate and tuffs. Among the sediments are a thick volcanic conglomerate, tuffaceous sandstone, sandstones, mudstones, a debris flow and a fossiliferous shale. The succession is about 240m thick, lavas accounting for approximately 150m.

ROUTE: The outward route from St Andrews is along the A91, crossing the late-glacial raised beach, to Guardbridge, thence by A919 through Leuchars to A92 as far as the fork at 438233. Thereafter follow an unclassified road north-west across the fluvioglacial sands and gravels of the Wormit Gap (Excursion 6). Cross the A914 at Links Wood Roundabout and continue on the unclassified road to a T-junction. At this junction turn right onto the B946 into Wormit. Two hundred metres north-east of Wormit post office, turn sharp left onto Bay Road

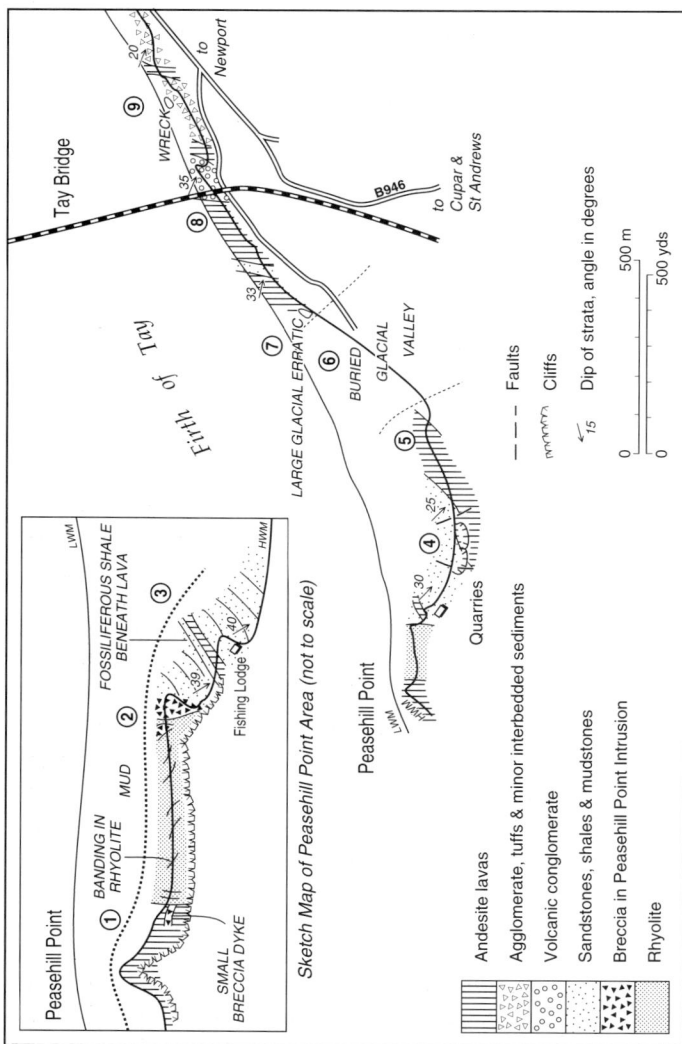

MAP 10: **Wormit coast.**

leading to the shore and passing beneath the Tay Railway Bridge. Dismount at the bay, send the bus back to the bridge and proceed west along the shore for a little less than 1km, past the small ruined fishing lodge to Peasehill Point (383258). The section can now be examined by walking back along the shore to the east as far as a point 400m ENE of the railway bridge.

1. Peasehill Point: lavas and margin of rhyolite

At Peasehill Point rather deeply weathered andesite lavas, of the Ochil Volcanic Formation, are exposed. They are grey and rotten at HWM, brown and fresher above this and cut by carbonate veins. Twenty-five metres east of the point the western margin of the Peasehill Point rhyolite is reached. The colour contrast between the grey andesite and the salmon pink rhyolite is pronounced. At the contact between the two is a thin (10cm) sheared layer. Next to this highly crushed layer, the rhyolite is whitened for about 30cm. The andesite is also altered, but in a much wider (5m) zone. Eastwards along the foot of the cliff the rhyolite is usually thinly flow banded. Notice too an 8cm thick red calcite vein 15m east of the contact. The banding, which is most obvious in weathered material and is accentuated by joints, is initially parallel to the margin of the intrusion at its western end but, after a gap, strikes NW–SE and dips steeply to the north-east. After about 90m the dip gradually changes through vertical until it dips to the SW at a point where the rhyolite abuts against a breccia which narrows into the cliff. The rhyolite varies in colour from pale grey through fawn to salmon pink. All varieties contain pale altered feldspar phenocrysts.

2. Peasehill Point: rhyolite and breccias

At the contact between the rhyolite and the breccia, notice just below HWM a small area of muddy sediments. Up the shore from this, the rhyolite appears to intrude the breccia and to roof it over, though the contacts are much obscured by vegetation. Eastwards the breccia is followed briefly by further exposures of rhyolite and then a second breccia. The banding in this rhyolite is abruptly cut off by the breccia. This breccia contains not only rhyolite blocks, but also blocks of sandstone and shale, particularly on the eastern side. Two possible explanations for the rhyolite can be offered. Harry (1956, p55) suggested that the rhyolite represented a vent intrusion, the breccia largely preceding the rhyolite. Alternatively, Armstrong *et al.* (1985, p36) suggest that the rhyolite is in fact a lava (as suggested by Geikie 1902, pp36–7), which may have lain within a small vent and have been faulted down into its present position from a higher level.

3. Fishing lodge: sediments, lava flow and fish band

Eastwards from the rhyolite a series of sediments can next be examined. Fifteen metres of yellow-brown gritty sandstones with mud pellets and cross bedding dip south-east and pass beneath 1.8m of dark grey, carbonaceous shales (exposed at the base of the cliff), which have yielded a considerable fish fauna: *Brachyacanthus, Ischnacanthus* and *Mesacanthus*, together with *Kampecaris, Pterygotus* and *Parka*, are recorded by Westoll (1950, p8), but a prolonged search is usually necessary to find even traces of these. Armstrong *et al.* (1985, p100) have suggested that this fauna is not of great stratigraphical significance. A short distance west of the fishing lodge, a decomposed lava overlies the shales which have been squeezed up into the base of the flow. The flow has locally been converted to white trap at its base, through reaction with the carbonaceous sediment, and the shales have been baked at the contact.

4. Old quarries: sandstones

For the next 180m there follow scattered small outcrops of flaggy and muddy sandstones, purple and grey in colour and occasionally coarse grained. This succession is about 60m thick and the sandstones were at one time worked in an old quarry 180m ESE of the fishing lodge, the remains of an old pier used for their export still being visible north of the quarry. In the quarry two small faults cut the sandstones, which are here tuffaceous. Near the top of the sandstones on the shore, small scale graded bedding can be seen just before they are succeeded by a thick sequence of lavas. Such sediments are typical of the Dundee Formation, lateral equivalent of the Ochil Volcanic Formation. Here the two are interbedded.

5. Andesite lavas

The next 275m of the shore section is composed of basic pyroxene-andesite lavas. These are rather deeply weathered and are greenish in colour on account of a high content of secondary chloritic minerals. Feldspar phenocrysts, though small, often demarcate flow lines in the rock. When fresh, such andesites contain pseudomorphs after olivine phenocrysts, plagioclase feldspar phenocrysts and rare augite phenocrysts all set in a groundmass of feldspar laths and tiny augites (Armstrong *et al.* 1985, p40). Red or purple staining is common and may be on the tops of some flows. Tops and bottoms of flows are not, however, conspicuous. The lavas are amygdaloidal, the amygdales being filled with chlorite, calcite, clear crystalline quartz or agate. At the footpath sign near the end of the concrete wall, the rocks at HWM show sediments washed into cracks in the lava top and red lateritic weathering.

6. Buried valley

The 0.5km gap in exposures which forms Wormit Bay is due to an overdeepened channel excavated by a branch of the Carse of Gowrie Glacier which flowed through the Wormit Gap to the south-east (see Excursion 6).

7. Glacial erratics, andesite lavas and sandstones

Beyond the bay a 3.5m long erratic block lies on the first lava outcrop. It is composed of Dalradian schistose grit brought from the Highlands by the Carse of Gowrie ice during the last glacial advance, as was a partly buried large block of dioritic or granodioritic rock a few metres away.

The lavas here are red or purple stained and are much altered andesites. They are highly amygdaloidal, chlorite being the commonest filling. Tops of flows can be seen from time to time picked out by lateritic weathering. As was the case further west, much sediment is trapped in the flows though there is little sediment between them. One hundred and forty metres north-east of the large erratic block, note the occurrence of 3.5m of chocolate-red, muddy, feldspathic sandstones which interrupt the lava sequence and form a band across the shore. The base of the sandstones, when exposed, may be seen to rest on an uneven surface of lava. The base of the overlying lava can be seen to have a sharp contact with the sediment, but in places seems to have sunk into the sediment below producing an irregular contact. The base of the lava is rubbly and locally vesicular. Kokelaar (1982) has described very similar phenomena from the base and top of andesite sills within the Old Red Sandstone volcanics in Ayrshire. The succession of flows continues for a further 180m to a point 75m from the bridge, with only two 30cm thick beds of red-brown micaceous mudstones to break the sequence. The rocks are similar to those described above. One of the flows, fresher than most, proved to be hypersthene-andesite when examined under the microscope.

8. Tay Bridge: conglomerate, sandstone, and lava

Seventy-five metres short of the bridge, a 1m thick green sandstone rests on the slaggy top of a flow. It is succeeded by a thick basic-lava volcanic conglomerate which extends to the remains of a pier built on a rocky point 75m beyond the Tay Bridge. The conglomerate comprises rounded blocks of andesite, often amygdaloidal and not infrequently containing feldspar phenocrysts up to 1cm long. Individual blocks average about 15cm in length with a maximum of about 30cm. The matrix is composed of fine lava debris and scattered through the conglomerate are impersistent lenses of cross-bedded sandstone or grit. Jointing in the rock cuts both pebbles and matrix, and calcite veins are common. Exposures, though continuous in the cliff, are best examined on the wave-washed surface at the pier mentioned above.

Six metres of tuffaceous sandstone or tuff, containing larger rock fragments towards the top, overlie the conglomerate at the ruined boathouse and remains of a jetty. Beyond this boathouse to the east are two thin autobrecciated lava flows which have a great deal of green sandstone mixed through them; so much, in fact, that it is difficult to recognise the rock as a lava at all. A little further to the east this is followed by volcanic conglomerate with both angular and rounded blocks of lava in a tuffaceous matrix. This is a rock in marked contrast to the volcanic conglomerate at the old pier where the blocks are well rounded. This conglomerate does not appear to be the product of a single eruptive episode however, for a 1m thick tuff or tuffaceous sandstone interrupts the conglomerate at one point and, beyond an iron pipe which crosses the shore, the fragments are very well rounded, a feature suggesting reworking of the material. A further pause in accumulation is indicated by a 2m thick band of lateritic tuff a few metres below the top of the conglomerate, 40m east of the pipe.

9. Wreck: volcanic-detrital sandstone and debris flow

The top of the conglomerate is marked by a 1m thick bed of green volcanic-detrital sandstone which crops out in the cliff opposite the wreck of a ship about 12m long, lying in the river mud and forming an obvious landmark except at high tide. This sandstone is overlain by further conglomerate and breccia with sharp angular blocks of andesite and rhyolite about 5cm in diameter in a reddish-brown matrix. The junction with the underlying sandstone is quite sharp. Upwards the blocks in the breccia increase in size up to about 45cm in diameter and rarely to 2m. In this rock have been found signs of plastic deformation of the fragments and some alignment of the smaller clasts. Recently such rocks have been interpreted as debris flow deposits, e.g. Armstrong *et al.* (1985, p.38). Blocks in this rock yielded the rhyodacite glass studied by Judd (1886). It is followed sharply upwards by a 4.5m thick lava flow containing amygdales filled with agate. Beyond this and still dipping south-east at 20°, volcanic conglomerates and tuffs, containing rhyolite fragments and blocks up to 2m across, continue for another 275m.

Return to the bus by retracing your steps westwards as far as the lifeboat house where a path ascends to the road, and return to St Andrews by retracing the outward route.

References

ARMSTRONG, M., PATERSON, I. B. and BROWNE, M. A. E., 1985. Geology of the Perth and Dundee districts. *Mem. Br. Geol. Surv.* Sheets 48W, 48E, 49.

GEIKIE, A., 1902. The Geology of Eastern Fife. *Mem. Geol. Surv. Scotland.*

HARRY, W. T., 1956. The Old Red Sandstone Lavas of the Western Sidlaw Hills. *Geol. Mag.* **93**, 43–56.

JUDD, J. W., 1886. Appendix to Durham, J., Volcanic rocks of the northeast of Fife. *Quart. Jour. Geol. Soc. Lond.* **42**, 418.

KOKELAAR, B. P., 1982. Fluidization of wet sediments during the emplacement and cooling of various igneous bodies. *Jour. Geol. Soc. Lond.* **139**, 21–33.

WESTOLL, T. S., 1950. The vertebrate-bearing strata of Scotland. *Rep. XVIIIth Int. Geol. Cong.* Part II, 5–21.

Excursion 6

St Fort–Leuchars (half day)

OS 1:50,000 map, sheet 59
GS 1:50,000 Sheets 48E, 49 drift and solid
Excursion map 11.

WALKING DISTANCE: 5km on farm and gravel pit roads.
Before visiting sand and gravel pits it is essential that permission to enter is first sought from the owners and that hard hats are worn. At the time of writing the owners are:

- St Fort Gravel Pit – Scottish Aggregates, Newport-on-Tay tel. (01382) 541659
- North Straiton Pit – Fife Sand and Gravel, Letham (Fife) tel. (01337) 810394
- St Michael's Pit – Tilcon Scotland Ltd., Leuchars tel. (01334) 839363

PURPOSE: To examine the fluvioglacial deposits of the Wormit Gap. These include the following: an esker, 4km long; other gravel ridges and mounds; kettle holes; three large gravel plateaus with surfaces at 33 to 40m OD; gravels and sands, probably reworked by the late-glacial sea and, associated with these, a series of late-glacial and postglacial raised beaches. They are exposed in an area 6.4km long by 4km wide extending south-east from Wormit to Leuchars, and are related to the retreat stages of the Late Devensian ice sheet. A Carboniferous tholeiite dyke is exposed at Newton Farm (Location 1).

141

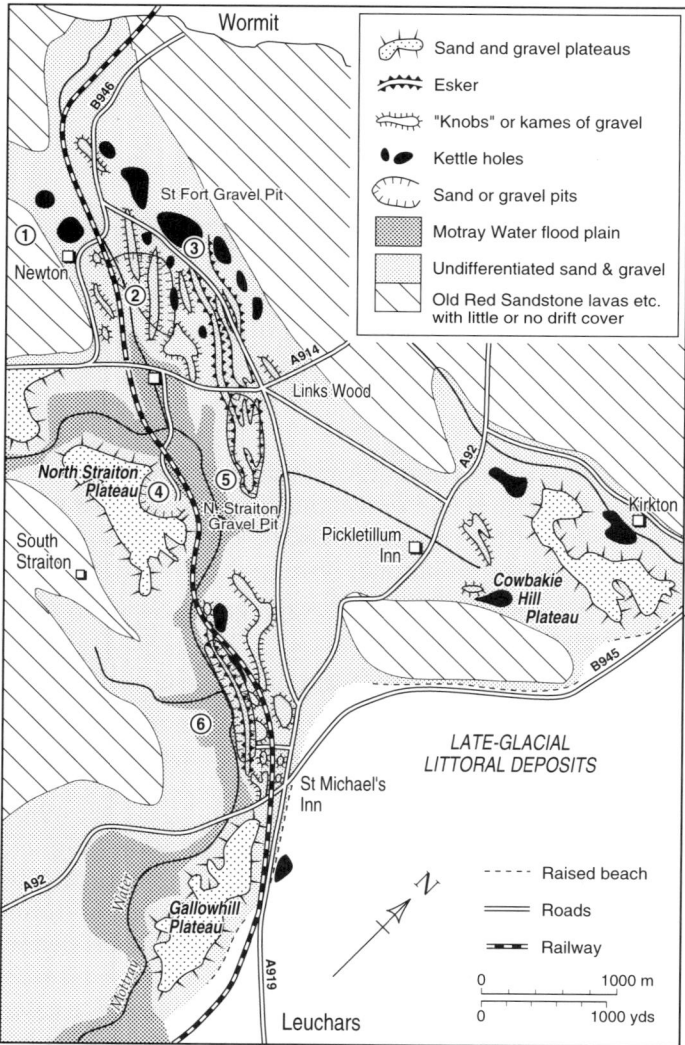

MAP 11: Glacial geology, St Fort - Leuchars.

ROUTE: By bus from St Andrews along A91 across postglacial raised beaches past Easter Kincaple to Guardbridge. Then by A919 through Leuchars, past St Michael's cross-roads to the fork at 438233; continue on the unclassified road to the round-about on A914. Follow A914 south-west for 1km before turn-ing right and following B946 to Newton Farm, 1km on the left.

1. Newton Hill: tholeiite dyke, andesite lavas, general view of fluvioglacial deposits

From the farm follow the track uphill. One hundred metres from the first gate, at the farm yard, are weathered exposures of the local Devonian, andesite lavas. Continue up the track to the third gate, some 600m from the farm yard. On the left is a 4m high cliff of columnar-jointed tholeiite in an E-W late-Carboniferous dyke. This dyke is better exposed in an old quarry further east and reached by contouring round the hill. There the dyke is 27m thick and the chilled northern margin can be examined at the east end of the quarry. The tholeiite is vesicular, fine- to medium-grained, mid grey in colour and very fresh. It displays horizontal columnar jointing. The coun-try rock comprises feldsparphyric, purple-weathering 'an-desite' and it is part of a flow within the thick Ochil Volcanic Group of Lower Devonian age, here dipping south-east at about 20°. These flows lie on the south-eastern limb of the Sidlaw Anticline. The north-western limb of the anticline can be seen across the Tay Estuary in the Sidlaw Hills beyond the Carse of Gowrie where similar lavas and interbedded sedi-ments dip at 20° NW. Note also Dundee Law, a volcanic plug of augite-porphyrite with its later development as a crag and tail, the tail to the east, comprising sediments which have been protected from the eastward moving ice by the porphyrite plug during the Devensian glaciation. Contrast the steep scarp slopes in the volcanics on Newton Hill with the gentle dip slopes, under cultivation, on Wormit Hill on the other side of the Wormit Gap.

Either the quarry or the road below afford a good view of

the fluvioglacial landforms of the Wormit Gap. Despite their steady disappearance as the fluvioglacial deposits are exploited for sand and gravel it is still possible to see good examples of kettle holes and the intervening steep-sided gravel ridges. At the St Fort Gravel Pit, clearly seen below, a borehole sunk in 1934 (Davidson, 1935) starting at 24m OD penetrated 40m of sand and gravel before entering andesites at 16m below OD. Thus the Wormit Gap with its thick fluvioglacial sequence overlies a deep channel presumably cut by ice during the Late Devensian glaciation, and analogous to the much deeper channel in the Tay going down to –68m OD below the road bridge. This area passes south-east in Links Wood into one of elongated gravel ridges analogous to eskers.

2. St Fort Gravel Pit: fluvioglacial gravels and sands in gravel ridges

Descend from the quarry, travel under the railway and turn east to enter the St Fort Gravel Pit where extraction of both sand and gravel from the gravel ridges takes place. Sand and gravel are interbedded in approximately equal proportions, but vary from 25 to 75 per cent of either at any time. The sands, usually reddish in colour, include darker beds seldom more than 5cm thick, dominated by igneous, mainly lava, fragments. Grain size ranges from fine to very coarse sand and sorting is usually poor. Cross bedding and ripple marks are widespread. Small scale faults occur and slumping too. Dips are up to about 20° and have been described by Rice (1961) as at right angles to the length of the ridges.

The gravels which are very poorly sorted range up to 1m in clast size in beds seldom more than a few metres thick. In places the gravels are clearly channel deposits. The clasts comprise: (1) metamorphic rocks including mica schists, quartzites, metamorphosed grits, vein quartz and epidiorite; (2) igneous rocks, predominantly andesites, usually feldsparphyric and vesicular, the vesicles often with chlorite or agate filling, agglomerates, tuffs and volcanic conglomerates; less common are diorite, porphyry and felsites; granite is rare;

(3) sediments, almost entirely sandstones which are usually red, sometimes with pebbles or mud flakes. Conglomerates also occur.

The volcanics and the red sandstones point clearly to a source area dominated by Devonian Old Red Sandstone rocks while the metamorphics are undoubtedly derived from the Scottish Highlands. Of major interest is the diorite, strongly suggestive of the Comrie Diorite 65km to the west. Much of the ice coming down the Tay and the Carse of Gowrie must have come from Strathearn to the west. Of note is the extreme rarity of rock types attributable to the Carboniferous.

3. Gravel ridges and kettle holes

On leaving the gravel pit turn north on B946 towards Wormit before turning east after 280m onto the unclassified road leading to the roundabout in Links Wood. Observe on both sides of the road kettle holes with steep-sided gravel ridges between them. Drainage is by downward percolation into the underlying sand and gravel and few of the kettle holes are wet. This topography continues to the roundabout. There turn south-west onto A914 and after 500m turn south-east onto the broad road which leads past the buried fuel tanks to North Straiton Pit.

4. North Straiton Pit: flat-topped plateau

Before going downhill from the main road to the North Straiton Pit observe the conspicuous flat-topped hill 0.5km to the south. The top of this plateau falls gently and evenly from 40m OD at its western end to 33m OD at its eastern end, a distance of about 1km. On all sides it slopes down steeply at about 20° from the almost flat top.

Park before reaching the Motray Water and walk under the railway bridge to reach the pit beyond. The face is between 15 and 20m high. In the lowest workings very well bedded fine and very fine sand and, in thin beds, silt occur. Cross bedding is usually low-angle; ripples, including climbing ripples, are present and sometimes evidence of slump scars 1–2m

145

deep are present. Lamination akin to varves has also been observed. Folds with 1–2m amplitude are sometimes exposed. The major part of the face displays cross bedding in units between 5 and 10 metres high; climbing ripples are common and thin beds with pebbles up to 10cm in diameter become commoner upwards. The topmost 1–2m of the face are pebbly and show distinct shallow channels. Cross bedding direction measurements over a number of localities in the topmost gravels round the plateau indicate currents flowing to the east and south-east.

There can be little doubt that these sediments have accumulated in a water-filled ice-bound crevasse, initially as fine sand and silt deposited under quiet 'glacio-lacustrine' conditions. As this body of water shallowed by silting up coarser material accumulated and both pebble beds and often large scale cross bedding formed. The climbing ripples point to periodic rapid influx of sediments. The pebbly final phase suggests streams crossing a very shallow body of water not dissimilar to a braided stream environment, but still ice bounded.

5. Links Wood: coarse gravels

Armstrong *et al.* (1985, pp75–8) have identified a central linear zone within the coarse esker gravels which pass from the St Fort Gravel Pit through Links Wood and on to the St Michael's Gravel Pit.

The Links Wood site though largely worked out can be reached by walking from the machinery site at North Straiton Pit north-east for 0.5km to the workings in what was formerly part of Links Wood. The sediments are similar to those in the St Fort Gravel Pit but are even coarser grained with boulders up to 1.5m diameter and less than 30 per cent sand. The boulders are of subangular and subrounded blocks, predominantly of Devonian volcanics and sediments, but still with metamorphic clasts, particularly of epidiorite.

6. St Michael's Gravel Pit: esker gravels and sands

Return to the bus and drive back to the main A914, crossing the Motray Water flood plain before turning north-east for 500m and at the roundabout turn south-east to St Michael's (2km). This stretch of road is quite flat and crosses what are now interpreted as late-glacial, marine, reworked sediments. These were formerly well exposed in a now filled-in sand pit at Brackmont Mill near St Michael's Inn (438224), where they displayed very abundant *Corophium* burrows and occasionally *Mcnocraterion* burrows (Buller and McManus, 1972). They are widely distributed at 24–27m OD between Leuchars and Tayport (Arbroath sheet 49, drift edition and Armstrong *et al.* 1985, p77). These sediments were deposited as the sea encroached on the steadily downwasting ice mass.

In the St Michael's Gravel Pit the 'central linear zone' gravels have been extensively worked where they formed a very clear esker ridge. Most of this has now been removed but that which remains forms a ridge 10m high consisting almost entirely of boulders of Devonian volcanics, volcaniclastics and sediments plus some metamorphics and up to 1m in diameter – less coarse than those at Links Wood (Location 5).

Level with this locality, but on the north-eastern side of the railway, is a pit in sands believed to be contemporary with the gravels to judge from their elevation. Ten metres of fine- to very fine-grained sometimes silty sands occur in crossbedded units up to 1m thick. The topmost 2m display pebble beds with clasts up to 25cm diameter. These are locally erosive into the sands beneath.

The remaining features of particular note in the area are the Gallowhill Plateau, on which the St Michael's Golf Course lies, and the Cowbakie Hill Plateau (see route map). These are closely analogous to the North Straiton Plateau (Location 4). Note that the Gallowhill Plateau descends to 27m at its southern end at Leuchars. Lastly one should note the major series of raised beaches extending eastward from St Michael's Inn to the coast and one in particular at Leuchars Lodge (445225) (Chisholm, 1966) in which a kettle hole was still occupied by ice until after this beach, at 17m OD, had been abandoned by the sea.

Interpretation

The interpretation of this complex set of late-glacial and post-glacial sediments has been set out with clarity in Armstrong *et al.* (1985, pp75–8). It is therefore only briefly summarised below.

The sediments have been brought by ice to the vicinity by the Late Devensian ice which lasted from 27,000–14,000 BP. As melting became greater than advance of ice from the west the ice became slow moving or stationary, thus these deposits show no evidence of over-riding or deformation by ice and the original glacially transported sediments were reworked by melt water.

Oldest are the 'central linear zone' gravels and sand ridges and eskers deposited subglacially, possibly under considerable hydrostatic pressure, by often swiftly flowing water.

As this ice continued to melt gaps or crevasses of considerable size developed in the ice, only to be filled with sediment, and the flat-topped plateaus are the products of this period. The height of these was most probably controlled by an englacial water table itself in turn probably controlled by sea level off shore to the east (see the Quaternary chapter for a regional picture).

Downwasting of the ice continued and parts of the area became ice free. Falling sea level at the same time prevented the destruction of most of the fluvioglacial landforms, but examples of the reworking of the fluvioglacial features are to be seen in the St Michael's area in particular and were formerly spectacularly displayed at Brackmont Mill. The southern end of the Gallowhill Plateau appears to have been eroded before sea level fell further and the last clear evidence of ice/marine water interplay is at the Leuchars Lodge kettle hole by which time sea level had fallen below 17m OD.

Return from St Michael's Gravel Pit by retracing the outward route noticing the steep northern end of the Gallowhill

Plateau analogous to the steep margins of the North Straiton Plateau (Location 4). Notice too the Leuchars Lodge kettle hole (445225) on the east side of the A919 road *en route* back to Leuchars and St Andrews.

References

ARMSTRONG M., PATERSON, I. B. and BROWN, M. A. E., 1985. Geology of the Perth and Dundee district. *Mem. Br. Geol. Surv.* Sheets 48W, 48E, 49.

BULLER, A. and McMANUS, J., 1972. *Corophium* burrows as environmental indicators of Quaternary estuarine sediments of Tayside. *Scott. Jour. Geol.*, **8**, 145–50.

CHISHOLM. J. I., 1966. An association of raised beaches with glacial deposits near Leuchars, Fife. *Bull. Geol. Surv. Gt. Br.* **24**, 163–74.

DAVIDSON. C. F. 1935. A boring at St Fort, Fifeshire. *Trans. Proc. Perth. Soc. Nat. Sci.* **9**, 167–9.

RICE, R. J. 1961. The glacial deposits at St Fort in north-east Fife: a re-examination. *Trans. Edinb. Geol. Soc.* **18**, 113–23.

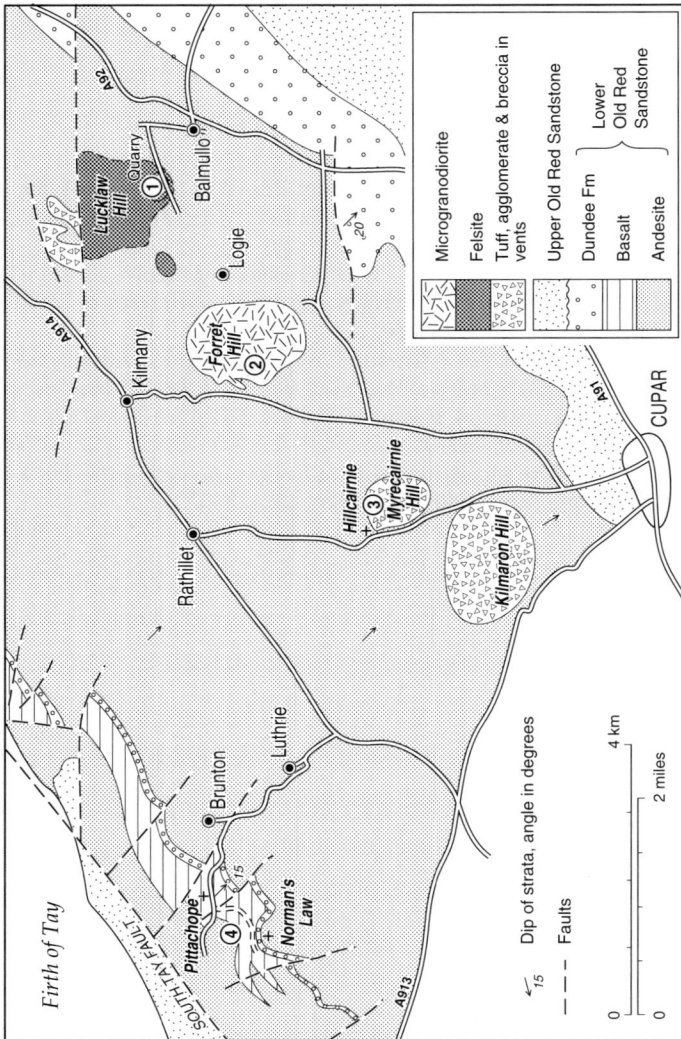

MAP 12: North Fife Hills.

150

Excursion 7

North Fife Hills
(half day)

OS 1:50,000 map, sheet 59
GS 1:50,000 sheets 48E, 49
Excursion map 12.

WALKING DISTANCE: Lucklaw Hill 0.5km, Forret Hill 1km, Myre-cairnie Hill 0.5km, Norman's Law 2.5km.

PURPOSE: The excursion is designed to examine (1) examples of the Lower Old Red Sandstone Ochil Volcanic Formation lavas of North Fife (2) the vents from which they may have been erupted and (3) a Lower Old Red Sandstone intrusion.

NOTES: Between St Andrews and Cupar, Lucklaw Hill forms a prominent mass 190m high, 3km north of the main A91 road. It is all the more conspicuous on account of a large quarry in its south face in which can be seen, even from several kilometres distance, the bright pink Lucklaw Hill Felsite. The felsite is regarded as an intrusion, possibly laccolithic, cutting the andesite and basalt lavas of the Lower Old Red Sandstone. The intrusion may have been a feeder for the adjacent lavas, though there are no rhyolite flows in the neighbourhood. A similar rock is exposed at Peacehill Point 4.5km to the north-west (Excursion 5).

A microgranodiorite intrusion, also cutting the lavas, forms most of Forret Hill 6km NNE of Cupar while Myrecairnie Hill, 4km north of Cupar, is composed of breccia and is tentatively believed to be part of a volcanic vent.

Norman's Law, 7km north-west of Cupar, comprises basalt and andesite lava flows and a few metres of grey, mica-rich, tuffaceous sandstones.

ROUTE: Proceed from St Andrews via the A91 to Guardbridge, then by the A919 for just over 1km before turning left onto an unclassified road leading past Leuchars Station, for 2km, to Balmullo. Cross the A92 at the village hall and after 200m turn north for 0.5km to Quarry Road which leads west to Lucklaw Quarry (419213) of the Fife Redstone Company. It is necessary to wear a safety helmet and to obtain permission to enter the quarry before proceeding.

1. Lucklaw Quarry: Lucklaw Hill Felsite

This forms a mass, of irregular outline, about 1.5km in diameter and, where exposed, with vertical margins. To the north, Armstrong *et al.* (1985, p43) have described a breccia of lava fragments in a felsitic matrix. This, they suggest, may be a down-faulted part of the roof of the felsite mass.

In the quarry the main workings are at a higher level, reached by a road leading west from inside the quarry entrance. At this higher level, the bulk of the felsite is an orange-pink, fine-grained rock with phenocrysts of orthoclase, plagioclase and rare biotite. In places the rock is banded and on weathering, purplish patches appear. Locally in the quarry, the felsite is pale grey in colour. Jointing, though conspicuous and often consistent for tens of metres in the quarry, does not appear to follow any systematic pattern, varying from vertical to horizontal. Veins of barytes up to several centimetres wide occur and are accompanied by very small quantities of green malachite. Other veins, 1–3cm across, are pale grey in colour and formed of crypto-crystalline quartz. Some joint faces reveal dark brown dendritic markings generally believed to be of manganese dioxide.

Microscopically the felsite comprises phenocrysts of ortho-clase and plagioclase in a fine-grained quartz and feldspar groundmass. The rock is described by Armstrong *et al.* (1985) as a rhyolite.

Return in the bus to the A92 and proceed south for 2km before turning west on an unclassified road signposted for Logie. At the T-junction after 2km, turn left for a further 2km before turning north on an unclassified road signposted to Kilmany. After 1.5km park the bus and walk 200m eastwards along a track to the quarry (388200), readily seen from the road, in the south end of Forret Hill.

2. Forret Hill Quarry: microgranodiorite intrusion

This mass is one of four intrusive bodies cutting the Lower Old Red Sandstone volcanics in the Cupar area of north-east Fife. It is 1.5km from north to south and 1km from east to west with the principal component a microgranodiorite. This rock type is well displayed in the quarry where it shows N–S vertical banding and jointing parallel to and close to the western margin of the body. The intrusion is regarded as a boss by Armstrong *et al.* (1985, p43), having vertical contacts with the country rocks.

On a fresh surface the rock is medium grey in colour with conspicuous feldspar phenocrysts up to 2mm long. A purplish hue appears with weathering and the rock occasionally displays a curious vesicular appearance. In thin section the rock can be seen to comprise plagioclase feldspar phenocrysts and pseudomorphs after pyroxene and magnetite in a quartz and alkali feldspar matrix (Armstrong *et al.* 1985).

The country rock andesite is exposed about 300m north of the quarry just beyond a swampy stream which crosses the rough track leading north and upwards from the quarry. The outcrop, which is about 30m above the fence, is of purplish, unusually hard, compact andesite which may have been baked in proximity to the microgranodiorite boss.

Continue north in the bus for 2km to Kilmany, turn west onto the A914 for 2km to Rathillet, then turn south for 2.5km to Hillcairnie Farm. Leave the bus and walk east through the farmyard (ask permission at the farm) and beyond for another

200m before a branch of the track runs 50m south to a quarry in a small wood (367185).

3. Myrecairnie Hill: volcanic neck

The Geological Survey map (sheet 48E) indicates that Myrecairnie Hill may be a volcanic neck though no boundaries to it are exposed. The rocks within it are displayed in the Hillcairnie Quarry and comprise pink and cream coloured felsite fragments together with purplish lava fragments. Armstrong *et al.* (1985, p37) record fragments up to 2m across and note silicification as common. Fragments are more usually a few centimetres across, but this is hard to gauge in the quarry on account of the lichen cover over much of the rock surface. Rock from the quarry seems, however, to have been used in the walls of the farm buildings and although now weathered, a better idea of the fragmental nature of the rock can be gained by examining blocks in the walls on the way back to the bus. Similar rocks crop out on Kilmaron Hill to the south-west and both hills are tentatively marked as volcanic necks on map sheet 48E. An alternative explanation offered by Armstrong *et al.* (1985, p37) is that they are altered volcaniclastic sediments. If volcanic necks they are likely sources for part of the very large thickness of Lower Old Red Sandstone volcanics in North Fife.

From Hillcairnie Farm return north to the A914 and turn south-west for 3km before turning north-west to Luthrie village and beyond. After 1.5km pass on the western side of the village of Brunton to just beyond Pittachope Farm and park at the entrance to a track (309209) leading south towards Norman's Law, the conspicuous hill 1km to the south.

4. Norman's Law: Old Red Sandstone lavas and interbedded sandstone

Armstrong *et al.* (1985, p21) have indicated that the Ochil Volcanic Formation of the Lower Old Red Sandstone reaches a thickness of 2400m in the area north-west of Cupar. This locality affords an opportunity of examining some of the lavas together with the landscape resulting from their folding to form the south-east limb of the Sidlaw Anticline. This anticline was later eroded to produce dip and scarp slopes over an extensive area in the North Fife Hills.

Within 50m up the track the first of three lava flows forming cliffs up to 3m high on the left of the road is reached. The lavas are purplish in colour, vesicular and often so weathered as to be crumbly. In some parts minor faulting can be seen. These flows have been identified as basalts. In the second flow examples of autobrecciation can be seen where blocks, up to 15cm across, of the lava's early formed vesicular crust have been incorporated into the still fluid parts of the flow forming a breccia. The vesicles contain dark green chlorite, calcite, zeolites and agates (in different areas); others are empty.

Follow the track uphill to the second cattle grid. Fifty metres before the grid the basalt is fresher and shows poor columnar jointing. Leave the track at the small gate on the south side beside the cattle grid. Notice the conspicuous scarp face of Norman's Law with prominent columnar jointing east of the summit. Walk 200m almost due south to the western corner of a pine plantation and join a rough track there running to the south-east. Follow this for 175m to a small gate on the eastern side where an ill defined path runs east for 120m along the fence to a quarry (311201). Here, medium-grained, well bedded grey sandstone dipping at 15° SE is exposed displaying minor cross bedding and conspicuous white mica on some bedding planes. About 4m of sandstones, overlain by 2m of dark grey siltstones, are exposed.

This sandstone and the overlying, unusually thick (130m) hypersthene–andesite lava can be traced for at least 15km along the strike of the North Fife Hills, with Norman's Law as the highest of several conspicuous summits formed from this flow. In each case the north-west face is marked by strong columnar jointing, characteristic of the middle of the flow, while to the south-east is a more gentle dip slope.

Return to the small gate and then follow the ridge to the summit of Norman's Law, at 285m the highest of the North

Fife Hills (305202). Notice the contrast, in this south-east dipping lava, between the scarp face, controlled by columnar jointing, and the dip slope in which jointing parallel to the top of the flow plays a major part. Hypersthene-andesites from the North Fife Hills have been described (Armstrong *et al.* 1985, p40) as containing hypersthene which is usually replaced by chlorite or serpentine. Other phenocrysts include rare altered olivine and rare plagioclase feldspar which also makes up the groundmass.

From the summit on a clear day, there are extensive views of the regional geology. To the east are the series of summits, with their dip and scarp slopes, formed of the same hypersthene–andesite flow as forms the summit of Norman's Law. To the south are the twin volcanic necks of the East and West Lomonds and also Largo Law, another volcanic neck. To the west lies the valley of the River Tay with Kinnoul Hill in Lower Old Red Sandstone lavas and sediments with, far beyond, the Dalradian rocks in Ben Vorlich. To the north lies the Firth of Tay and the Carse of Gowrie where Upper Old Red Sandstone and Carboniferous sediments have been downfaulted between the North and South Tay Faults. They are overlain by late-glacial clays and the postglacial Carse Clays. Beyond are the Lower Old Red Sandstone volcanics of the Sidlaw Hills, dipping to the north-west into Strathmore on the north-western limb of the Sidlaw Anticline. Beyond these, the Grampian Highlands, composed of older Dalradian metamorphic rocks, are seen.

The summit shows the remains of a first millenium BC hill fort with a series of 'defensive enclosures' (Walker and Ritchie 1987, p167).

Descend the western end of the ridge past good columnar jointing in the lava before turning north to reach the track where it enters the plantation north-west of Norman's Law. One of the basalt flows beneath the sandstone is well exposed at the roadside 75m west of the point where the track enters the plantation. Such basalts usually contain olivine, now replaced, and plagioclase feldspar phenocrysts in a groundmass of plagioclase feldspar and tiny augite crystals. Walk east back down the track to rejoin the bus.

Take the bus back through Luthrie to rejoin the A914 and follow this for 2km south-west to the crossroads with the A913. Take the A913 for 6km south-east to Cupar and from there the A91 back to St Andrews.

References

ARMSTRONG, M., PATERSON, I. E. and BROWNE, M. A. E., 1985. Geology of the Perth and Dundee district. *Mem. Br. Geol. Surv.* Sheets 48W, 48E, 49.

WALKER, E. and RITCHIE, G., 1987. *Exploring Scotland's heritage, Fife and Tayside.* Royal Commission on the ancient and historical monuments of Scotland. HMSO, Edinburgh.

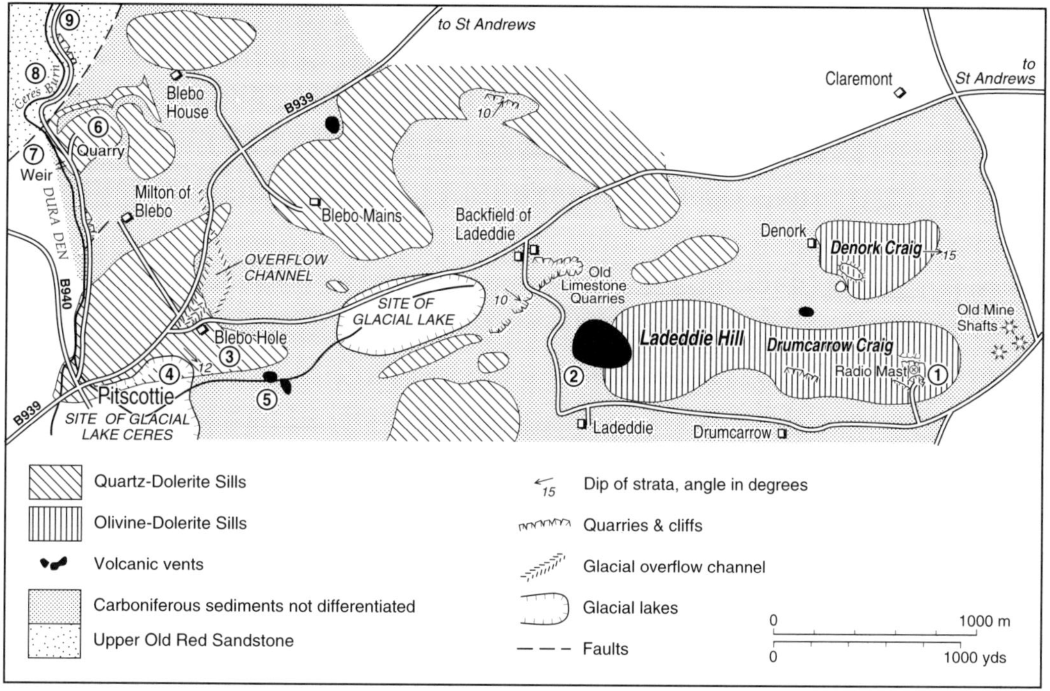

MAP 13: Drumcarrow - Dura Den.

Legend:

- Quartz-Dolerite Sills
- Olivine-Dolerite Sills
- Volcanic vents
- Carboniferous sediments not differentiated
- Upper Old Red Sandstone
- 15 Dip of strata, angle in degrees
- Quarries & cliffs
- Glacial overflow channel
- Glacial lakes
- – – – Faults

Labels on map: to St Andrews, Claremont, Denork Craig, Old Mine Shafts, Radio Masts, Drumcarrow Craig, Drumcarrow, Ladeddie Hill, Ladeddie, Denork, Old Limestone Quarries, Backfield of Ladeddie, SITE OF GLACIAL LAKE, Blebo Mains, B939, Blebo-Hole, OVERFLOW CHANNEL, Blebo House, Milton of Blebo, Quarry, Pitscottie, SITE OF GLACIAL LAKE CERES, DURA DEN, B940, Weir, to St Andrews

1000 m
1000 yds

158

Excursion 8

Drumcarrow and Dura Den (half day)

OS 1:50,000 map, sheet 59
GS One-inch, 1:50,000, sheets 41, 49
Excursion map 13.

WALKING DISTANCE: 3km of track and rough road.

PURPOSE: The main objects of this excursion are to examine: (1) some of the Upper Old Red Sandstone sediments in Dura Den; (2) Lower Carboniferous sediments at about the level of the Lower Ardross Limestone (see Table V); (3) the Drumcarrow Olivine-Dolerite Sill; (4) the Blebo Hole Quartz-Dolerite Sill and its contact effects upon the adjacent shales; (5) two small volcanic necks within Kinninmonth Den.

Owing to the complex structures and lack of exposures, the Carboniferous rocks have not been subdivided on the Excursion map.

ROUTE: Leave St Andrews on the B939 from the West Port and fork left at the University playing fields 1km further on. Continue past Craigtoun Park to Claremont crossroads (461146) and there turn left (south then south-west) for 1.5km along an unclassified road before forking right (west) for 120m to the track leading up to a quarry on the south face of Drumcarrow Craig. The top of the hill carries a radio mast and a television relay mast.

1. Drumcarrow Craig: olivine-dolerite sill

The quarry is cut into the Drumcarrow Olivine-Dolerite Sill which is 1.5km long from east to west and up to 450m wide from north to south. The majority of exposures of the sill show a well developed columnar jointing while small scarps on the hillside are due to an almost horizontal jointing at right angles to the columnar jointing. In the quarry itself, columnar jointing is prominent and large joint planes normal to the columns (parallel to the intrusion surface) dip steeply southwards. Towards the northern margin of the sill these joint planes are inclined gently north, while at the southern margin they are steeply inclined south. This sill has been classified (Forsyth and Chisholm 1977, p141) as a non-ophitic olivine-dolerite, one of a small group in East Fife. In the quarry the rock has a conchoidal fracture and yellow olivines can often be seen on fresh surfaces. Old mining records indicate that the southern side of the sill is strongly transgressive to the local sediments (Forsyth and Chisholm 1977, p142).

Ascend to the summit of Drumcarrow Craig, 90m north of the quarry, for a view of the surrounding geology. The olivine-dolerite sill of Denork, also showing good columnar jointing, lies 400m north-west where it is well exposed in Denork Craig. The low ground to the east is scarred by small tips and shafts from old ironstone workings centred on Denhead and dating from the 19th century. The ironstone, in the Lower Limestone Formation, lies in a faulted NE–SW trending syncline. Apart from the sills, exposure in this area is generally poor. On Drumcarrow Craig east–west glacial striae may be noted on some of the less deeply weathered crags of olivine-dolerite.

2. Ladeddie: volcanic vent

Return to the bus and continue west for 1.5km. The western end of the Drumcarrow Sill is cut by an olivine-basalt plug and agglomerate filled vent, just north of Ladeddie Farm. Old quarries in the field are situated where, on a much smaller scale than that of the sill, columnar jointing occurs in the fine-grained basalt.

3. Blebo Hole: glacial lakes, drainage channel, quartz-dolerite sill

Continue to the T-junction at Backfield of Ladeddie then turn westwards across the bed of a former postglacial lake. The high ground to the north and south of the lake bed is capped by quartz-dolerite sills which form bluffs and crags in several places. At Blebo Hole (423134), just over 1.5km west of Backfield of Ladeddie, park at the old farm building and follow the grass track south for 90m to a quarry in a steep west-facing scarp. To the west is a wide expanse of level ground extending almost to the village of Ceres. This is probably where a mass of dead ice melted forming Glacial Lake Ceres. What was a drainage channel from this lake, now dry, passes the old farm buildings at Blebo Hole, continues north for 400m, crossing the B939, runs past Blebo House and ultimately joins the Ceres Burn near Kemback. If this interpretation is correct, the present gorge of the Ceres Burn in Dura Den must be a postglacial feature. The burn in the gorge of Kinninmonth Den most probably drained the post-glacial lake to the west of Backfield of Ladeddie.

In the quarry, note the rich yellow-brown soil and subsoil

FIGURE 7: The flat fields are the site of a former glacial lake near Ceres, seen from Blebo Hole, Pitscottie.

passing down into poorly jointed, spheroidally weathering quartz-dolerite. The rock is coarse grained and contains numerous pink segregation patches. The form of weathering is perhaps the most obvious difference between this quartz-dolerite and the olivine-dolerite of Drumcarrow. In addition, iron pyrites is nearly always visible in the hand specimen of the quartz-dolerite, but is rarely seen in the olivine-dolerite (Irving 1929). The segregation patches are notably poor in ferromagnesian minerals.

4. Blebo Hole Marine Band

Ninety metres south of the quarry, just before the wall which passes in front of the quarry reaches the burn, at the base of a small tree, there is a small outcrop of grey shales dipping to the south-east at 12°: the Blebo Hole Marine Band (Forsyth and Chisholm 1977, p48). They are hard and have presumably been thermally metamorphosed by the quartz-dolerite sill. This sill forms the scarp 45m to the north-east. In the baked shales the fossils have been recrystallised and are now very conspicuous, occurring either as white crystalline calcite with a ferruginous stain, or, if the specimen has been long exposed to weathering, as moulds of the original shells, the calcite having been dissolved out. These shales are believed to be at approximately the same horizon as the shales lying just above the Lower Ardross Limestone. The fauna includes *Productus, Aviculopecten, Nuculana, Orthoceras, Straparollus, Fenestella, Lithostrotion* and crinoid ossicles (Craig and Balsillie 1912, p12). The marine band is also exposed on the opposite side of the burn, though here it is not as fossiliferous.

5. Blebo Hole: volcanic vents

If time is available, walk up the north bank of the burn past the previously examined quartz-dolerite sill. In the burn are several small outcrops of Carboniferous rocks lying above the baked shales. One hundred and seventy metres upstream from the shales, notice a white trap dyke exposed in the bed of the burn where it follows an S-bend. Three hundred metres upstream are exposures of a small vent 45m long and mainly tuff filled, the tuff containing a good deal of little-indurated shale (Craig 1912, p84). This rock is similar to that in many other vents in East Fife, but the two dykes that traverse it are relatively unusual (the most obvious of these is exposed in the burn just upstream from a tree which has been blown over and now lies on the fence). The groundmass of the dykes is hard, fine grained and highly altered, consisting of a paste of chlorite and calcite. In it are set xenocrysts of anorthoclase up to 5cm long, clear and colourless when fresh and yellow or brown when weathered. Large biotite xenocrysts are also common, hornblende xenocrysts rare. Fragments of sandstone and shale are also present, the latter usually surrounded by calcite, while pieces of dark grey glass also occur. A second vent filled with a similar tuff is cut through by the stream 55m further east, but no dykes are exposed in it. Its eastern contact with the country rock can be seen where shales bend abruptly downwards into the vent, a common feature of such vents owing to drag at the close of eruption when material collapsed back into the vent.

Return to the bus and 90m west of Blebo Hole, join the B939 for 400m to Pitscottie crossroads. Note the quartz-dolerite sill exposed on the right, 50m short of the crossroads. Turn right at the crossroads and immediately right again down Dura Den. For the first 550m the road crosses a quartz-dolerite sill which is exposed in the Ceres Burn. Eight hundred metres from the crossroads a NNE–SSW fault down-throwing on the east brings in 15m of horizontal, cross-bedded, creamy sandstone belonging to the Sandy Craig Beds of the Lower Carboniferous (Forsyth and Chisholm 1977, p12). Dismount at the weir at Grove Cottage (416142) and send the bus on for just over 1km to a small car park past a row of houses and opposite the village hall (416151).

6. Dura Den, Blebo Quarry: quartz-dolerite sill

At the roadside north from the weir, another quartz-dolerite sill is split by about 6m of sandstone which extends east for

400m. The upper leaf of the sill, now rather altered, may be examined in Blebo Quarry, about 110m up a narrow path leading north-east from the weir where the path turns sharply right. It is coarse grained and contains pink segregation veins. Now walk northwards along the main road. The sandstone bed which splits the sill crops out on the eastern side of the road, but the contacts with the lower leaf of the sill are not exposed. The first outcrops of the sill are, however, fine grained.

7. Dura Den Fault and upturned beds

Two hundred and seventy-five metres from the weir, just before a sharp right hand bend in the road, the lower leaf of the sill can be seen upturned against a fault and, beneath it, carbonaceous sandstones with a small coal seam are exposed. Fault drag has increased the dip of these beds to 40° to the south-east in an area of otherwise almost horizontal strata.

8. Dura Den: fossil fish locality

Follow the road downhill until it meets the Ceres Burn again 450m below the weir. Sandstones in the bed of the burn at this point belong to the Dura Den Formation of the Stratheden Group (see Table III), formerly simply referred to as the Upper Old Red Sandstone. They form part of the famous Dura Den fossil fish locality. The Dura Den Formation here comprises red, green and cream siltstones alternating with cream coloured sandstones, often with ripple cross bedding (Chisholm and Dean 1974, p19). Polygonal mud cracks occur too. The locality was extensively quarried by British Museum collectors (Woodward 1915) and seems to have been completely worked out. Commonest among the fish was *Holoptychius flemingi* but several other genera were also present and House *et al.* (1977, p74) have correlated these beds with the Upper Devonian (Famennian) beds at Clashbenny Quarry on the northern side of the Tay and with other localities in Scotland and Belgium. At the present day sandstones above the old mill lade on the western bank of the burn yield only

PLATE 2: *Holoptychius* sp. from the Dura Den Formation, Stratheden Group, Upper Old Red Sandstone. Excursion 8. Location 8. This specimen has been carefully cleaned for display. (Pocket knife 7cm long) (Photo N. Mackie).

isolated scales of *Holoptychius*. Attridge (1956) found a new locality higher in the succession that has yielded *Holoptychius*, but no further discoveries of this kind have been made. No fossils have been recorded from the eastern bank of the burn to date.

9. Dura Den: Dura Den Formation (Upper Old Red Sandstone)

Higher beds of the Dura Den Formation form a high cliff to the east of the road and can be examined 180 to 275m downstream before reaching a row of cottages and a telephone box between the road and the burn. There is a rough scramble to reach the base of the exposures in the cliff. In the cliff there are 20m of fine- to very fine-grained, slightly feldspathic sandstones which are in the main plane-bedded, soft, poorly consolidated, and yellow-brown to cream coloured. Some cross bedding occurs and is generally low angled in sets of around 15cm. Also displayed are ripple marks. These are generally 2.5–5.0cm in amplitude and asymmetrical. Climbing ripples are rare. Impersistent, thin (2.5cm), coarser beds of up to granule grade occur as do much harder 10cm thick nodular calcareous horizons and red-stained silty sandstones. Local erosion surfaces show undercutting. These sandstones appear to be mainly aeolian in origin. Three metres up the cliff and exposed on the underside of a shelf is a mud-flake breccia. This suggests that waterlaid as well as aeolian sediments are present (Hall and Chisholm 1987, p204). The nodular calcareous beds here and those at the village hall suggest incipient calcareous soil or 'calcrete' development analogous to that at Bishop Hill (Excursion 17) and fairly widespread in rocks of this age elsewhere in Scotland.

Now continue down to the waterfall just past the village hall. Here the sandstone again shows low angle cross bedding (but in sets up to 1.5m thick). The sandstone is fine grained and has numerous veins and nodules of calcite.

Rejoin the bus and continue down Dura Den past the old stone bridge over the River Eden (416161) and continue east for 2.5km towards Strathkinness.

On the slopes of Knock Hill there are several old overgrown quarries on both sides of the road. These are located in both the Pittenweem Beds and, above, the Sandy Craig Beds of the Strathclyde Group of the Carboniferous (Forsyth and Chisholm 1977, pp42–4) and were formerly worked for building stone, much of which can be seen in St Andrews where Geikie (1902, p346) remarked on its poor resistance to weathering. Continue eastwards noticing the fine view on the northern side of the road across the Eden Estuary to the Leuchars–Tentsmuir expanse of blown sand. On entering St Andrews, note the University Playing Fields and Observatory standing on a fluvioglacial terrace sloping down eastwards from 30m to 25m in a distance of around 1.5km (Cullingford and Smith 1966, p37).

References

ATTRIDGE, J., 1956. A gigantic *Holoptychius* from Dura Den. *Nature Lond.*, **177**, 232–3.

CHISHOLM, J. I. and DEAN, J. M., 1974. The Upper Old Red Sandstone of Fife and Kinross: a fluviatile sequence with evidence of marine incursion. *Scot. Jour. Geol.*, **10**, 1–30.

CRAIG, R. M., 1912. Additions to the volcanic geology of East Fife. *Trans. Edinb. Geol. Soc.* **10**, 83–9.

———————— and BALSILLIE, D., 1912. The Carboniferous rocks and fossils in the neighbourhood of Pitscottie, Fifeshire. *Trans. Edinb. Geol. Soc.*, **10**, 10–24.

CULLINGFORD, R. A. and SMITH, D. E., 1966. Late-glacial shorelines in Eastern Fife. *Trans. Inst. Brit. Geogr.*, **39**, 31–51.

FORSYTH, I. H. and CHISHOLM, J. I., 1977. The Geology of East Fife. *Mem. Geol. Surv. Gt. Brit.*

GEIKIE, A. 1902. The geology of Eastern Fife. *Mem. Geol. Surv. Scotld.*

HALL, I. H. S. and CHISHOLM, J. I., 1987. Aeolian sediments in the late Devonian of the Scottish Midland Valley. *Scot. Jour. Geol.*, **23**, 203–8.

HOUSE, M. R. *et al.*, 1977. A correlation of Devonian rocks of the British Isles. *Geol. Soc. Lond. Spec. Rep.*, No. 7.

IRVING, J., 1929. The Carboniferous igneous intrusions of North-eastern Fifeshire. *Unpublished Ph.D. thesis, St Andrews University.*

WOODWARD, A. S., 1915. Preliminary report on the fossil fish from Dura Den. *Rep. Br. Ass. Advmt. Sci.,* 1914, 122–4.

Kinkell Braes, St Andrews (half day)

OS 1:50,000, sheet 59
GS 1:50,000, sheet 49
Excursion map 14.

WALKING DISTANCE: 1.5km on path, 1.5km on rocky shore.

PURPOSE: This is a straightforward elementary excursion introducing a variety of geological phenomena. These include: (1) common sedimentary rocks; sandstone and shale in particular and also mudstone and limestone; (2) common features of sedimentary rocks such as cross bedding, ripple marks and joints – these are widespread; (3) fossils which occur in certain beds only, mainly those deposited under marine conditions; (4) folding and faulting which are well displayed; (5) a landslip.

The beds belong to the Strathclyde Group of the Carboniferous System (see Table IV) and show many signs of having accumulated under shallow, freshwater or deltaic conditions with occasional marine incursions. Igneous rocks, though present, are rare and highly altered. Among much more recent features are the late-glacial and postglacial raised beaches, the Maiden Rock sea stack and erratic blocks of dolerite left behind by ice during the Quaternary glaciation of the region. At the present day, a still younger beach is forming complete with its wavecut platform cutting across rocks of varying hardness on the shore.

169

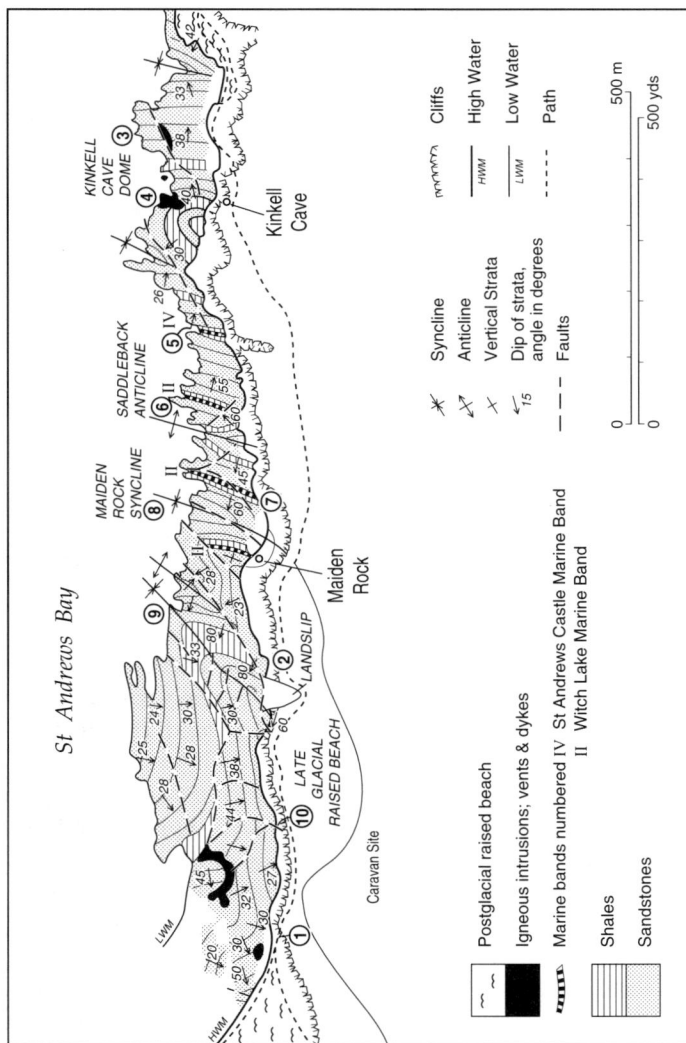

St Andrews Bay

KINKELL
CAVE
DOME ③
④
Kinkell
Cave

SADDLEBACK
ANTICLINE

MAIDEN
ROCK
SYNCLINE

Maiden
Rock

LANDSLIP

LATE
GLACIAL
RAISED BEACH

Caravan Site

Postglacial raised beach

Igneous intrusions; vents & dykes

Marine bands numbered IV St Andrews Castle Marine Band
                          II Witch Lake Marine Band

Shales

Sandstones

⚹ Syncline

⚹ Anticline

✕ Vertical Strata

↙15 Dip of strata,
   angle in degrees

– – – Faults

〜〜〜〜 Cliffs

—HWM— High Water

—LWM— Low Water

– – – – Path

0 _____ 500 m
0 _____ 500 yds

MAP 14: Kinkell Braes, St Andrews.

Since many of the exposures are in the intertidal zone, the excursion should be undertaken as near to low tide as possible.

ROUTE: From the public car park at the Albany Park Flats (near the Gatty Marine Laboratory) at the East Sands, St Andrews, follow the shore path which runs southwards for 370m across a low raised beach (with the Leisure Centre to the south-west) and then climbs up the cliff at the beginning of the Kinkell Braes.

1. Kinkell Braes cliff path: raised beaches

Halfway up the steep part of the path observe the loose sand and mud that have fallen down from above. This is sediment from the late-glacial raised beach at the top of the cliff on which the lower part of the caravan park is situated. The sediments contain shell fragments not very different from those on the present day beach. Contrast the weathered and grassed over state of the cliff at the back of this raised beach with the fresher cliff at the back of the younger, postglacial raised beach and that at the back of the present day beach.

2. Landslip

Follow the coast path along the top of the cliff for 200m. At the eastern end of the caravan site is a slow-moving, active landslip. It can best be viewed from the eastern side from the top of the shore cliff. Much of the material slipping towards the beach is shale and the toe of the slip extends below high water mark (HWM) on the beach, where it is eroded by the sea thus promoting further slipping. Large blocks of sandstone, soil and bushes can be seen on the head of the slip and, from time to time, the curved fault planes on which they move.

In a more general view of the shore below, a U-shaped outcrop can be discerned with the top of the U opening to the south-west, i.e. obliquely into the cliff. This is a syncline: i.e. a trough-shaped fold in the rocks with the beds on the northern side dipping towards the cliff at 30° and those on

FIGURE 8: Saddleback Anticline, Kinkell Braes, St Andrews. The ridges are composed of sandstones, one of which, on the right limb of the fold, is cut by a small fault.

the south-eastern side dipping into the trough much more steeply at 70°–80°. Further east, the strata (beds) can be seen to strike (general trend or run of the beds) parallel to the cliff until at the Maiden Rock they are folded into another syncline. Follow the path eastwards until at a point 180m beyond the Maiden Rock, where the path turns south up a slight hill, the rocks on the shore can be seen to be folded into an anticline (the Saddleback Anticline): i.e. they are arched upwards and dip outwards on either side at 30°–60°. Beyond the anticline the beds again dip east, until after 270m they are sharply folded into a small syncline near LWM. Immediately east of this, a semicircular dome-like structure in the rocks can be seen from the cliff top – the Kinkell Cave Dome.

Now follow the path down the cliff to the postglacial raised beach and to the shore itself. The remaining part of the excursion can now be carried out by walking back along the shore to the East Sands.

3. *Sandstones, shales, a fault and a white trap dyke*

At this locality the strike of the beds is N–S and the dip is 25°–40° E. The sandstones are yellow to buff in colour and vary from massive beds showing trough cross bedding to thin-bedded sandstones, often with ripple marking and a tendency to break into blocks along joints. The finer-grained, softer, grey, shaly beds have been eroded out by the sea to leave trenches on the shore. These are often partially filled with boulders or sand thus making it difficult to examine the shales. Halfway down the shore notice an ENE–WSW trending fault which cuts the sandstones and displaces the seaward outcrops east. This is a dip fault, i.e. it runs parallel to the direction of dip of the beds that it cuts. By examining closely the line of the fault, a creamy-buff rock can be seen in the fault plane and pinching out locally along it. This is 'white trap': an alteration product of a basalt dyke that has been intruded along the line of the fault. The original dark coloured, basic igneous rock has undergone considerable chemical change owing to the incorporation of much CO_2, produced from the carbon of the country rock. This originates in Carboniferous plant debris buried in the sediment when it was laid down. The resulting carbonate-rich 'white trap' bears little resemblance to basalt (elsewhere a gradual transition from basalt to white trap may be seen, e.g. Excursion 15, Location 3). The fault is difficult to follow through grey shales and sandy shales, but can be located again to the west where it cuts a thick sandstone before dying out. When this sandstone is followed by eye up the cliff, Kinkell Cave can be clearly seen.

4. *Kinkell Cave Dome and Vent; small syncline*

Beyond the ridge on the shore formed by the thick sandstone is the Kinkell Cave Dome in which the beds dip outwards to the north, east and west. The arching of the beds over the dome can be seen in the cliff behind. Low on the shore and cutting the thick sandstone, a small volcanic vent, filled with white trap, is exposed (Balsillie 1920 (b), p81). The vent measures 45m by 60m and has an irregular outline with a narrow tongue extending south into the country rock. A separate part of the vent on the east of the sandstone ridge and lying a short distance offshore is filled with volcanic ash or tuff containing fragments of sediment. Follow the thick sandstone westwards from the cave round the dome and into the sharp syncline where, partly because of movement along a small fault, it stands up in a V-shaped ridge of particularly well jointed sandstone.

5. *St Andrews Castle Marine Band*

Now walk west across a series of east-dipping sandstones until after 100m a deep little cove runs into the cliff. This has been excavated by the sea along a band of richly fossiliferous, dark grey shales containing red-weathering ironstone nodules and thin limestone bands. The detailed section in the cove is as follows:

	metres
Massive sandstones	—
Sandy shales	0.9
Thin bedded yellow-brown sandstones	2.0
Shale with *Naiadites* becoming sandy upwards	0.9
Fossiliferous limestone	0.5
Fossiliferous shale with ironstone bands	6.0
Coaly shale	0.3
Massive sandstones	—

The fossiliferous limestone and shale form the St Andrews Castle Marine Band (Forsyth and Chisholm 1977, pp37–41), the fossils including '*Productus*', *Aviculopecten*, *Myalina*, *Naiadites*, *Euphemites*, fish teeth and abundant crinoid ossicles. The small fault which cuts the sharp syncline low on the shore at Locality 3 can be seen cutting this marine band a short distance below HWM, but the throw here is only a few metres and a short distance further west the fault dies out. Now walk westwards across 60m of east-dipping strata consisting mainly of sandstones, but which have been eroded out to form trenches, now partly filled with boulders.

6. Witch Lake Marine Band; Saddleback Anticline; faulting

At the western end of these sandstones and dipping beneath them, is a second marine band, the Witch Lake Marine Band, probably equivalent to the Pittenweem Marine Band in Table V. It outcrops on both sides of the Saddleback Anticline and the detailed successions are listed below (after Kirk 1925 and Forsyth and Chisholm 1977, p39):

		metres	
		West	East
(a)	Thin-bedded sandstone and shale	3.0	2.0
(b)	Fossiliferous shale with ironstone nodules	2.3	2.5
(c)	Crinoidal limestone	0.15	0.3
(d)	Shale becoming sandy towards base	2.1	2.4
(e)	Sandstone	—	—

Units (b) and (c) are the Witch Lake Marine Band

Brachiopods, bivalves and gastropods can be collected from the shales and also weather out on the surface of the limestone. Crinoid ossicles and rare crinoid cups occur in the limestone. When the Witch Lake Marine Band is followed up the shore towards HWM on the eastern side of the Saddleback Anticline the outcrop is cut off by a NW–SE fault which dies out to the north-west. Drag of the beds against the fault can be clearly seen and indicates a sinistral movement, i.e., the beds on the side opposite the observer have moved to the left. Forty-five metres west from the Witch Lake Marine Band examine the axis of the Saddleback Anticline. The beds dip at 45° E on one side and 45°–60° W on the other, while at the 'nose' of the anticline they dip at 10°–25° N. This is the plunge of the fold. Notice that individual beds can be followed round the nose of the fold from one side to the other and on looking into the cliff behind, some of the thick sandstone beds can be seen arching over the fold. The Saddleback Anticline is cut by another NW–SE fault which, though it has its greatest displacement on the western side (or limb), cuts right across the middle of the fold. This fault also has a sinistral displacement and since it displaces the fold axis, can be described as a tear fault.

7. Witch Lake Marine Band; faulting

The succession already examined on the east side of the anticline is now repeated on the west side with the brown-weathering sandstone again forming a prominent ridge 50m west of the fold axis. The Witch Lake Marine Band also reappears, but is obscured at HWM by a strike fault, i.e. one that runs approximately parallel to the strike of the bed that it cuts. The line of this fault is marked by a gap in the brown sandstone ridge, the sandstone having been shattered by the faulting and thus made more susceptible to erosion by the sea.

8. Maiden Rock Syncline; the Maiden Rock

The next 75m of the shore are occupied by strata, higher in the succession than the Witch Lake Marine Band, which are folded into the Maiden Rock Syncline. On the eastern side of this fold the beds dip at 60°–80° towards the axis of the fold. The western limb has been steepened so much, however, that the beds have been overturned and now dip at angles of up to 130° towards the east, i.e. they dip west at angles as low as 50°. In this fold too, individual beds can be followed round the nose of the fold where the plunge is about 35°N in approximately the same direction as the Saddleback Anticline. The Witch Lake Marine Band reappears once more a few metres west of the Maiden Rock, but is poorly exposed at HWM where its course is marked by a boulder and sand filled trench.

At this point notice the prominent Maiden Rock, an old sea stack standing on the remnant of the postglacial raised beach. It dates from the time when the sea level stood some 4m higher than now and consists of vertically dipping sandstone, bounded on the east and west by bedding planes and on the north and south by minor fault planes. These planes of weakness were preferentially eroded by the sea thus leaving a sandstone pillar as an erosional remnant.

A short distance north-west of the Maiden Rock, the Maiden Rock Syncline is abruptly cut off by a NE–SW fault, the course of which is marked on the beach by a trench filled with boulders and bottomed by shattered rock. For 180m to the west, sandstones dip north at 25° until another NE–SW striking fault is reached. Notice also the gap in the cliffs formed by a landslip. The first fault is believed to have a considerable displacement because the familiar marker horizon, the Witch Lake Marine Band, is not exposed again before the Castle at St Andrews, 1.5km to the north-west. In the shatter belt marking the second of these faults, Balsillie (1920a, p76) records the presence of volcanic tuff intruded up the fault plane. This tuff is now almost totally obscured by boulders.

9. Tightly folded syncline; accommodation faulting

Beyond this fault sandstones low on the shore are folded into a small anticline which is succeeded to the west by the prominent syncline seen earlier from the top of the cliff. Notice that the centre of the syncline is occupied by thick, trough cross-bedded sandstones underlain by grey shales with ironstone nodules and that preferential erosion of the shales by the sea has picked out the structure. The shales are in turn underlain by a thick series of sandstones. In the syncline, dips are as high as 80° on the eastern limb but, no more than 30° on the northern limb while the fold plunges south-west at about 15° A series of small WNW trending faults, apparently associated with the intense folding, cuts the thick sandstones in the centre of the syncline. The shales on the other hand have deformed plastically round the more rigid sandstones. A fault at the foot of the cliff and running parallel to the shore, separates the syncline from the beds in the cliff which dip towards the WNW at 60°. Details of the structure are not known, but this is probably another tear fault.

Notice the toe of the landslip, mentioned at Locality 2, where it spills onto the shore. Much of it is made up of soft shale. It has been moving slowly since 1982.

10. Fault drag; volcanic vents

From the landslip to the East Sands, a distance of about 0.5km, the shore is largely occupied by sandstones. In detail a number of faults break up the succession and one of these is exposed in the cliff about 180m east of the point where the path begins to climb the cliff. This fault, which is apparently a low angle one, downthrows to the east and drag of the beds as they approach the fault plane can readily be seen from the shore below.

Secreted in cracks and faults in the rocks of this part of the shore are a number of small masses of tuffs, but usually these can only be located after intensive search. The most prominent is U-shaped with sandstone striking N–S inside the U and sandstone striking E–W outside the U. This particular volcanic vent has been eroded by the sea to form a trench about 3m wide near LWM, but the tuff is completely covered by sand and cannot be seen without excavating. A white trap dyke, seldom more than 60cm thick, runs westwards from this vent but is largely covered by sand.

Well developed ripple marks can be seen in thin-bedded sandstone outcropping in the cliff below the path where it rises from the postglacial raised beach. Towards the western end of exposures on the shore, the strike of the beds gradually swings round to the NW–SE, the general dip remaining about 35° towards the south or south-west. Just before the end of exposures at HWM, the beds are cut by another fault trending NNE. Across this fault there is a pronounced change in the strike of the beds. Sandstones outcrop on both sides and the displacement is unknown.

References

BALSILLIE, D., 1920, a. Descriptions of some volcanic vents near St Andrews. *Trans. Edinb. Geol. Soc.*, **11**, 69–80.

———, 1920, b. Descriptions of some new volcanic vents in East Fife. *Trans. Edinb. Geol. Soc.*, **11**, 81–5.

FORSYTH, I. H. and CHISHOLM, J. I., 1977. The geology of East Fife. *Mem. Geol. Surv. G.B.*

KIRK, S. R., 1925. A Coast section in the Calciferous Sandstone Series of Eastern Fife. *Unpublished St Andrews University Ph.D. thesis.*

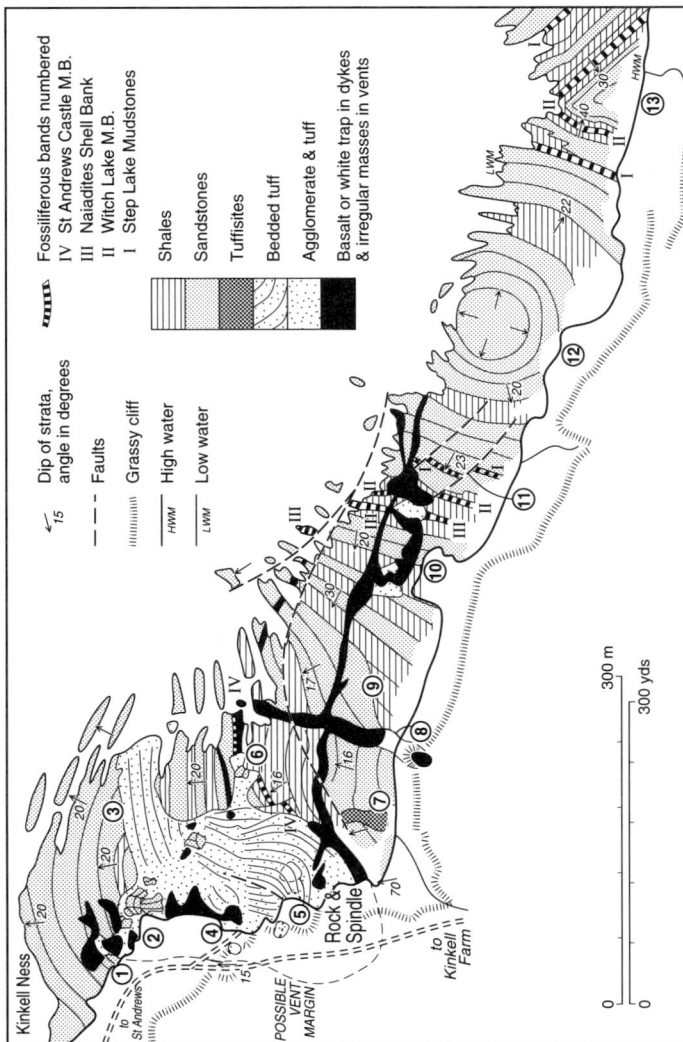

MAP 15: Rock and Spindle - Craigduff, St Andrews.

Kinkell Ness

Fossiliferous bands numbered
IV St Andrews Castle M.B.
III Naiadites Shell Bank
II Witch Lake M.B.
I Step Lake Mudstones

Shales
Sandstones
Tuffisites
Bedded tuff
Agglomerate & tuff
Basalt or white trap in dykes & irregular masses in vents

Dip of strata, angle in degrees
Faults
Grassy cliff
High water
Low water

Rock & Spindle

to St Andrews

POSSIBLE VENT MARGIN

to Kinkell Farm

300 m
300 yds

180

Excursion 10

Rock and Spindle, St Andrews (half day)

OS 1:50,000 map, sheet 59
GS 1:50,000, sheet 49
Excursion map 15.

WALKING DISTANCE: 3.25km on grassy paths, 3km on rocky shore.

PURPOSE: The main objects of this excursion are to examine: (1) Lower Carboniferous rocks belonging to the Strathclyde Group at the horizon of the Witch Lake Marine Band (probably equivalent to the Pittenweem Marine Band of Table V); (2) a series of exceptionally well exposed Carboniferous volcanic vents or necks which permits the study of the mechanism of emplacement and stages of development believed to be typical of such vents in Fife; (3) folding and faulting affecting both the sedimentary and igneous rocks. Much information on both the detail and the broad setting of these rocks is to be found in Forsyth and Chisholm (1977).

Since this is a coastal excursion it should be carried out at as near low tide as possible.

ROUTE: On foot for 2.5km from the East Sands at St Andrews, following the coastal path as for Excursion 9 until a branch leaves the cliff top and descends to the postglacial raised beach at the far end of the Kinkell Braes, 1.3km from the East Sands, and thereafter continuing on this path for a further 550m to

Kinkell Ness (537157) at the western end of the shore section to be examined.

1. *Kinkell Ness Vent: country rock*

Kinkell Ness is composed of sandstones dipping to the north at 20° and lying between 120m and 150m stratigraphically above the local Strathclyde Group marker horizon, the Witch Lake Marine Band. The sandstones are cross-bedded, cream-coloured and contain plant fragments. Small veins of better cemented sandstones stand out on weathered surfaces.

2. *Kinkell Ness Vent: northern end*

A prominent low stack near HWM marks the northern limit of the Kinkell Ness Vent, sometimes known as the Rock and Spindle Vent. The stack is formed of basalt packed with sedimentary xenoliths (especially sandstone) round the margins and xenoliths of igneous material towards the centre. Parts of the host rock near the centre are suffeciently free of xenoliths to show its basaltic nature.

Before going on to describe the exposures within the Kinkell Ness Vent, the stages in its history as worked out by Kirk (1925b), are listed below:

(i) The initial volcanic outburst consisting mainly of gas discharge which produced the vent or pipe.

(ii) Repeated pyroclastic activity, the formation of the bedded tuff cone and subsequent collapse of the bedded tuff into the vent.

(iii) Emplacement of vent intrusions consisting of basalt generally packed with xenoliths, but locally relatively free of these.

(iv) Emplacement of basalt dykes which may extend outside the vent.

The northern stack clearly comprises rocks from the third stage in this sequence, which are separated from the main part of the vent by rocks belonging to the first two stages. The second stage rocks should be examined 45m south-east of the stack where they consist of: (a) blocks of sandstone up to 6m across with quite random orientation, the bedding lying in attitudes varying from vertical to horizontal; (b) agglomerate or tuff composed of fragments, usually a few cms across, of shales, sandstones, 'rotten' lava and white trap; (c) a matrix which is fine grained and seems to be simply an aggregate of much smaller fragments of the same rocks. Kirk (1925b) records sand grains and tiny rock fragments in the tuff. The relationship of this tuff to the sandstone blocks serves to illustrate the mechanism of intrusion at an early stage in the history of the vent. The fine grained tuff can be seen to penetrate into cracks and joints in the sandstone while in several places the bedding of the sandstone is destroyed, the resultant rock being a mixture of sandstone and fine tuff. Such a relationship suggests very strongly that gas action has broken down the cohesion of the sandstone and, to judge from the amount of fine shaly dust in the tuff, it seems probable that the shales of the country rocks have been broken down in a similar fashion.

3. Kinkell Ness Vent margin

On the northern side of the large eastern extension of the vent which runs down to LWM, the margin of the main vent is exposed. The contact with the sandstones of the country rocks is sharp and the latter may have been slightly baked. At this point the vent is occupied by poorly bedded tuff, often quite coarse, which, on erosion by the sea, gives rise to a more even surface than the sandstones. Elsewhere the contacts are seldom exposed.

4. Kinkell Ness Vent: xenolithic basalt and ash

Eighteen metres south of the large sandstone blocks within the vent at Locality 2, examine another mass of basalt, also packed with xenoliths and forming the next stack close to HWM. This has, at its southern end, a well exposed contact with the bedded tuff. The basalt is altered for up to 10cm from the contact and has taken on a pale green colour. Xenoliths

FIGURE 9: The Rock and Spindle, Kinkell Braes, St Andrews. The radiating spokes of the 'spindle' are of columnar-jointed basalt; the tall 'rock' is of xenolithic basalt. The cliffs and stack in the background are composed of bedded tuff, all lying within the Rock and Spindle Vent.

in the basalt consist of older basalt, sandstone, shale, mudstone, coal and finely bedded tuff. This exposure lies just north of the point where the road from Kinkell Farm reaches the shore.

5. Kinkell Ness Vent: bedded tuff

Bedded tuff is well seen on the wave cut platform just in front of the point where the road reaches the shore. Bedded tuff also forms the next massive stack, 6m high, some 24m NNE of the Rock and Spindle. In the massive stack, graded bedding, with the finer beds repeated every 15cm or so, can be examined (especially on the seaward face). Larger blocks are scattered through the tuff and occasionally cross bedding can be discerned. Among the larger blocks, dark masses of lava and dove-grey limestones (which weather almost white) are conspicuous. The latter sometimes contain the coral *Lithostrotion* and crinoid ossicles. Although the bedding in the tuff is approximately vertical throughout this area, the direction of the top of the succession can be ascertained from the graded bedding and also from the rare signs of large fragments having dropped into the soft tuff and disrupted the bedding. The steep dip of the tuff is undoubtedly a result of collapse of the bedded tuff (presumably in a cone) into the vent after eruption as envisaged by Francis (1962, p56, fig 7) and Forsyth and Chisholm (1977, p200). The arcuate strike of the tuff suggests a volcanic centre somewhere near the end of the road where it reaches the beach. Beyond an east–west fault the tuff strikes north-east, but the bedding disappears when followed further.

6. Kinkell Ness Vent: dyke, faulting, country rock, St Andrews Castle Marine Band, the Rock and Spindle

The eastern side of the vent 110m north-east of the Rock and Spindle stack is packed with large masses of sediment. These consist mainly of sandstone, but include blocks of limestone up to 2m long. The orientation of the large blocks is random, but the bedding in the tuff is very approximately parallel to the vent margin.

Another smaller extension of the vent lies beyond this area to the east and a dyke, now altered to white trap, extends still further to the east along the outcrop of the St Andrews Castle Marine Band. This band consists principally of marine shales, bearing abundant ironstone nodules and thin limestone bands at most a few centimetres thick. *Naiadites* is commonly present in the shales. By following this marine band westwards extensive small scale faulting can be discerned. The outcrop of the St Andrews Castle Marine Band is then interrupted by the vent for 25m before it can once more be picked up and followed through a further series of minor faults. Beyond this it terminates against the vent margin, 70m to the south-west. At this point, the vent margin is steep and the peripheral xenolithic basalt intrusion stands up as much as 3m above the soft shales of the country rock. Cutting both this intrusion and the shales is a basalt dyke of stage iv that can be followed westwards through the bedded tuff to the Rock and Spindle. The Rock and Spindle is a basalt mass of stage iii, in the main richly xenolithic, especially in the tall thin 'Rock'. The 'Spindle' on the other hand, is relatively free of xenoliths and shows good columnar jointing, the radial pattern of which indicates a pipe-like form. No clear boundary between this rock and the xenolithic rock can be discerned and they are probably part of the same intrusion. The contact between the intrusion and the bedded tuff can be followed readily on the western side of the stack while at the northern side, it is possible to determine that dyke emplacement occurred very late in the history of the vent since a 0.5m basalt dyke cuts both the bedded tuff and the xenolithic basalt.

At the top of the beach, just outside the margin of the Kinkell Ness Vent, there is evidence for the collapse of not only the bedded tuff, but also some country rock into the vent. For a few metres outside the probable position of the margin, the country rocks dip at 70° into the vent in marked contrast to its dip of 16°–18° to the north a short distance further south.

7. 'Explosion fissure'

Twenty-five metres south-east from the vent margin notice a gap about 12m wide in the thick sandstone, running due north and narrowing in that direction. The narrow northern end of the gap merits careful examination. It is the locus of a stage i vent (Kirk 1925b) occupied by randomly arranged blocks of shale and sandstone in a shaly matrix, quite free from igneous material, a rock often referred to as tuffisite. It is what Geikie (1902, p210) called an explosion fissure, presumably formed by the upward passage of hot gases. The country rocks along its margins are slightly indurated and shattered.

Twenty-five metres north of this vent the dyke cutting the south-east margin of the Kinkell Ness Neck is again met in the shales beneath the St Andrews Castle Marine Band. If time permits, its complex course, controlled by faulting, should be followed eastwards for 90m to the next vent.

8. Dyke-like vent with white trap

This vent, 155m long by 12m wide, is dyke-like in form and has been emplaced along the line of a pre-existing minor fault. It is cut by the small dyke referred to above and also by a later tear fault, which displaces the vent dextrally by 12m. Except at its two ends, the vent has been eroded out by the sea to form a trough on the beach. Near its centre the vent is composed principally of basalt, but towards its margins it is packed with xenoliths of now thermally metamorphosed country rock. Towards HWM the basalt gradually alters to white trap and contains many calcite veins. Locally along the margins, relics of tuff from early stages of emplacement remain.

9. Soft sediment slump scar; white trap dyke

The sandstone cut by the southern part of this vent overlies a series of shales and an examination of the base of the sandstone between 4.5m and 45m east from the vent reveals very clearly a surface cut into the beds beneath. At least 3m of shaly beds are successively cut out as the surface is traced down the beach to the north-east. Above the cross-cutting surface a few centimetres of thin-bedded sandstone are everywhere succeeded by 10–15cm of grey shales with ironstone nodules. Only above this is there a thick cross-bedded sandstone with clear channels cut into the underlying beds.

By analogy with the modern sediments of the Mississippi delta the cross-cutting surface is likely to mark the site of a slump scar in the unconsolidated sediments on which a mass of sediment has moved downslope. Only later have the thin-bedded sandstone and the shales been deposited on the surface of the slump scar.

Forty-five metres downshore, the sediments are cut by the dyke mentioned above. Inconspicuously exposed, the dyke is altered to white trap, in both the shale and the sandstone. It continues east to the Craigduff Vents and beyond before disappearing under the sea.

10. *The Craigduff Vents: Naiadites shell bed*

The irregular form of the Craigduff Vents can be seen from the map. They are occupied in the main by xenolithic basalt, but also contain a little tuff. The country rocks have been baked and some of the adjacent sandstone now stands up as quartzite stacks on the shore. Inspection of the map reveals that a mass of sandstone is almost entirely cut off within the vents though maintaining its original dip and strike. At the eastern end of this mass, the vent margin can be clearly seen. Rafts of sediment are present in the basalt and are thoroughly baked. Some basalt, relatively free from xenoliths, occurs in the eastern part of this series of vents; this rock has developed horizontal columnar jointing and forms a small stack. Within the central area of the series, a patch of tuff has been eroded by the sea to leave a gap in the shore crags. On the shore opposite this examine one of the most striking exposures in the Strathclyde Group sediments, a yellow or buff weathering *Naiadites* shell bed which can be seen to thin from 2m to 0.7m within a few metres when traced up the shore. This is the *Myalina* limestone referred to by Geikie (1902, p213) and Kirk (1925a, p368). A few centimetres beneath this limestone is a 30cm coal seam resting on a seatearth. Both the limestone and the coal are abruptly truncated by the vents and fragments of limestone have been found within the vents. The thermal metamorphism adjacent to the vent is very pronounced in the coal, a natural coke having been produced in the seam for about 30cm outwards from the contact, and the vesicles in the coal are filled with white calcite. This rock is, however, no longer exposed.

11. *The Witch Lake Marine Band and a tear fault*

Twenty metres east of the *Naiadites* shell bed, the Witch Lake Marine Band outcrops as a 15cm limestone band in about 6m of dark grey shale. This sequence of richly fossiliferous beds, containing brachiopods, bivalves and gastropods, is underlain

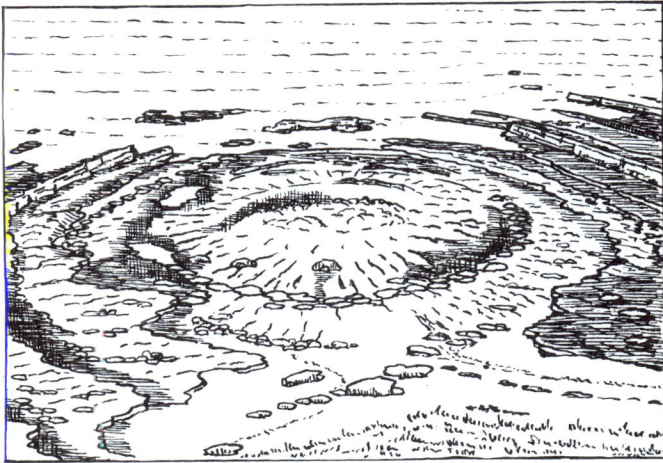

FIGURE 10: The Craigduff Dome seen from the cliffs of the
Kinkell Braes, St Andrews. In this structure the
Carboniferous sandstones dip outwards in all directions
at 18-20°.

by sandstones then shales. These richly fossiliferous beds are
then followed 40m further east by the Step Lake Mudstones,
only about 3m thick and here consisting of dark grey mudstones with calcareous nodules. The outcrop of these two
bands is interrupted by a sinistral tear fault with an 18m
displacement and striking NNW.

12. The Craigduff Dome

The strata intervening between the Step Lake Mudstones and
the centre of the Craigduff Dome (543154) comprise 55m of
mainly yellow-weathering, grey sandstones with a 9m shale
interrupting the continuity. Ironstone nodules occur in the
shale which is not, however, very fossiliferous. The sea has
eroded the shale away round much of the dome, thus serving
to accentuate this structure. The dome itself is remarkably
symmetrical, the dips being 18°–20° outwards in all directions.

At its centre the sandstones have been eroded by the sea into grotesque shapes, partly controlled by cementing along rami-fying calcite veins.

13. The Craigduff Syncline: penecontemporaneous erosion

One hundred and eighty metres east of the Craigduff Dome, the Step Lake Mudstones and the Witch Lake Marine Band may be examined once more on both sides of the Craigduff Syncline. Between them occurs an example of a sandstone with an erosional base cut into the beds beneath. In the beds overlying the Step Lake Mudstones on the eastern side of the syncline, there are only 3m of sandstone overlying 25m of shale. On the western side, however, there is a 12m sandstone immediately above the Step Lake Mudstones. Follow the base of this sandstone downshore. It can be seen to rise steadily from the shales which are readily recognised here on account of the 50cm thick hematitically weathering limestone within them. The sandstone is cross-bedded and truncates successive beds of shale when followed up the shore. Rip-up clasts of shale are present in the basal part of the sandstone. At the top of this thick sandstone, a 60cm band of muddy *Naiadites* limestone is locally present on the western side of the syn-cline, but is absent elsewhere. Small faults which provided accommodation during the folding of the syncline can be observed in the sandstone above the Witch Lake Marine Band, especially where puckering is present in the shale.

Return to St Andrews by the coastal path via Kinkell Braes.

References

FORSYTH, I. H. and CHISHOLM, J. I., 1977. The geology of East Fife. *Mem. Geol. Surv. G.B.*

FRANCIS, E. H., 1962. Volcanic neck emplacement and subsidence structures at Dunbar, Southeast Scotland. *Trans. Roy. Soc. Edinb.*, **65**, 41–58.

GEIKIE, A., 1902. The geology of Eastern Fife. *Mem. Geol. Surv. Scot.*

KIRK, S. R., 1925a. The geology of the coast between Kinkell Ness and Kingask, Fifeshire. *Trans. Edinb. Geol. Soc.*, **11**, 366–82.

——————, 1925b. A coast section in the Calciferous Sandstone Series of Eastern Fife. *Unpublished St Andrews University Ph.D. thesis.*

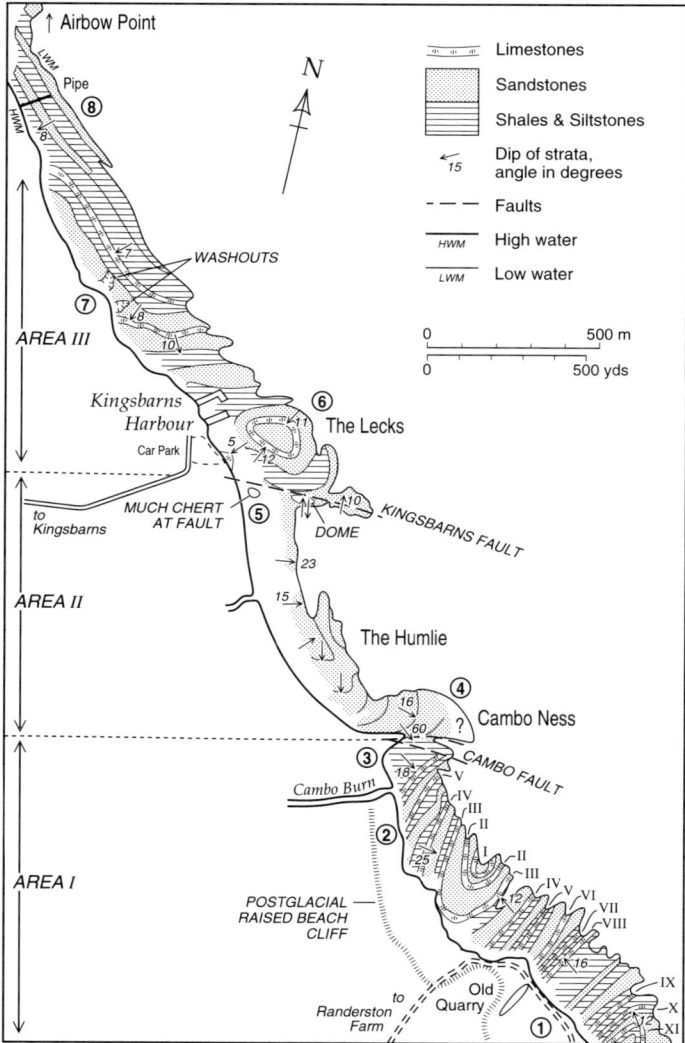

Limestones

Sandstones

Shales & Siltstones

Dip of strata,
angle in degrees
15

Faults

HWM   High water

LWM   Low water

0   500 m
0   500 yds

↑ Airbow Point

N

LWM
Pipe
⑧
HWM
8

WASHOUTS

⑦
8
10

AREA III

Kingsbarns
Harbour   ⑥   The Lecks
Car Park
5
11
12

MUCH CHERT   ⑤
AT FAULT   DOME
10
KINGSBARNS FAULT

to
Kingsbarns

AREA II

23

15

The Humlie

④
16
Cambo Ness

60   ?

③
CAMBO FAULT
18   V

Cambo Burn
IV
III
②
II
25   I   II
III
IV V VI
VII
VIII
12

AREA I

POSTGLACIAL
RAISED BEACH
CLIFF

16

IX

to
Randerston
Farm

Old
Quarry
①
X
XI
12

MAP 16: **Kingsbarns - Randerston.**

192

Kingsbarns–Randerston (half day)

OS 1:50,000, sheet 59
GS One-inch, 1:50,000, sheets 41 and 49
Excursion map 16.

WALKING DISTANCE: 5.5km of farm road, rocky and sandy beach.

PURPOSE: The main objects of this excursion are to examine:
(1) Carboniferous rocks known to lie very low in the Strathclyde Group succession (Table IV) of East Fife (including the
well known Randerston Limestones: see Table V); (2) strata
belonging to the Balcomie Beds of the Inverclyde Group.

NOTES: The shore section is divided into three areas by faults
and the relationships between the areas is unknown. Only the
rocks in Area I have been correlated with others in Fife.
Forsyth and Chisholm (1968, p75) reported a marked similarity between the Area I rocks and those of the Anstruther
borehole. Neves *et al.* (1973), however, disputed this on palynological evidence. Both groups, however, place them in the
Anstruther Beds of the Strathclyde Group. The rocks in Area
II are the oldest in the section and can be correlated with the
Inverclyde Group Balcomie Beds of Fife Ness. The age of these
beds is however, unclear. Forsyth and Chisholm (1977, p8)
placed them in the Upper Old Red Sandstone and Browne
(1980) correlated them with the Downie's Loup Sandstone of
the Inverclyde Group in the Stirling area. The sequence of
Area III is thought to belong to the lower part of the

Anstruther Beds of the Strathclyde Group (Forsyth and Chisholm 1977, p20), but no detailed correlation exists.

ROUTE: The outward route from St Andrews is by A917 travelling south-east through Kingsbarns (9.5km) to the entrance to Randerston Farm (602101). Dismount, send the bus back to Kingsbarns Harbour and walk down the farm road (seek permission at the farm) to the shore. Follow the track eastwards, noting *en route* the good cliff feature at the back of the postglacial raised beach and pass the old quarry where Limestone VII of the Randerston Limestones was once worked, but which is now filled in. Alternatively, take the bus to Kingsbarns Harbour and walk south-east along the coast to this point.

Four hundred and fifty metres south-east from the point where the farm road reaches the coast, there is an anticline on the shore (615110), the rocks in the centre of which comprise the lowest part of the succession in Area I. The succession within Area I can now be examined by walking along the shore to the north-west, studying the section through this anticline and the syncline which follows it to the north-west.

For 600m the strata exposed on the shore display an alternation of sandstones and shales with subsidiary siltstones, sandy shales, thin limestones, now dolomite (the Randerston Limestones) and coals, the last being seldom exposed. The simplified succession tabulated below is based largely on that in the East Fife Memoir (Geikie 1902, pp123–6) after Kirkby (1901). The general inclination of the beds is 10°–20°NW as far as the synclinal axis.

The beds were numbered individually by Kirkby and his numbers for the limestones are given below to the left of the limestone's name.

	metres
TOP	
Cross-bedded sandstones	3.5
Gap with boulders	1.8
1. LIMESTONE I; grey, hematite-stained with calcite veins	0.35

Sandstones	1.2
Shales with sandy ribs	1.8
5 LIMESTONE II; Yellow-buff weathering, grey, thin	
bedded	0.6
Shale	0.9
Thick sandstone, thin bedded at top, cross bedded	
and slumped below; muddy at base	6.7
Shale with ironstone nodules	0.3
11. LIMESTONE III; thin bedded, very shelly with	
Naiadites and ostracods	1.2
Shales with ironstones ⎫	
LIMESTONE IIIa ⎬ seldom exposed, gap	6.2
Coal and fireclay ⎭	
Thick sandstone with cross bedding and slumping	
from NNE thin bedded at base	6.7
18. Gap with poor exposures of LIMESTONE IV (30cm)	
in shales and resting on coal	2.4
Shales with 60cm sandstone at top	4.2
Sandstone, yellowish, cross bedded above, thin	
bedded below	3.5
Shale	1.8
25. LIMESTONE V; in two leaves, shelly with shale	
between; contains *Sanguinolites, Schizodus,*	
Aviculopecten	0.9
Shale with siltstone and sandy partings	5.5
Thick-bedded sandstone with *Stigmaria*; well jointed	
with stepped, angular appearance	6.0
Shales with ironstones near top	2.7
37. LIMESTONE VI; compact, grey, with shelly bands	
with high spired gastropod *Donaldina* (this unit is	
found 6m in front of the next scarp)	0.4
Shales, calcareous at top	1.8
Sandstones, cross bedded, thin bedded with coaly	
partings, clay at base (this unit forms a scarp)	6.0
Grey shales	1.8
42. LIMESTONE VII; irregularly bedded with	
Camarotoechia (the only limestone with articulate	
brachiopods), poorly exposed	0.4
Shale with sandy bands	3.0
Hard calcareous sandstone	0.3

Shale with ironstone and cementstone bands	3.0
Sandstone	0.6
Shale with ironstone and sandstone bands	8.0
64. LIMESTONE VIII; with hematite stain at LWM; flat topped	0.45
Alternating shale and sandstone in beds of up to 1.5m	9.0
Thin-bedded sandstone	4.5
Massive yellow sandstone, top 6m with slumping and cross bedding; forms flat surface on the shore	12.2
Shale with sandstone bands	4.2
82. LIMESTONE IX; stromatolitic with algal balls and hematite stain, ostracods*	0.22
Seatearth with calcareous concretions	0.45
Sandstone, grey and yellow	1.2
Shale with sandy partings	1.2
88. LIMESTONE X; shelly with hematite stain	0.3
Shales with 60cm yellow, sandy rib	5.5
94. LIMESTONE XI; buff, hematitic weathering, uneven surface	0.6
Marl; grey with calcareous concretions	3.0
97. Sandstones, massive below, shaly above	6.0
BASE	

1. The Randerston Limestones succession; SE side of the syncline

Within this succession the following features may be noted: (a) the thick, well bedded sandstones form prominent ridges on the shore, but, where strongly cross-bedded or contorted, they are planed off remarkably flat as wave cut benches without dip or scarp slopes; (b) Limestone VII is the only limestone with articulate brachiopods; (c) Limestone IX is a most striking rock with its hematite stain and stromatolites* (see plate 3); (d) Limestone VI contains the high-spired gastropod *Donaldina;* (e) Limestone V, a short distance east of the farm road, is very fossiliferous and the top surface, usually covered in

---

* Please do not collect from the outcrop of this bed. There is abundant loose material on the beach.

PLATE 3: Algal stromatolites and oncolites in Limestone IX of the Randerston Limestones, Strathclyde Group, Carboniferous; Randerston. Excursion 11, Location 1. (Photo J. A. Weir).

seaweed, yields *Schizodus, Sanguinolites, Naiadites, Murchisonia* etc; (f) the shales beneath Limestone III have been preferentially eroded to form a sandy bay which is clearly discernible from the road when approaching the shore; (g) Limestone III is shaly, thin bedded and packed with *Schizodus*. Note that the highest beds in the syncline can be seen only at low tide.

2. *The Randerston Limestones succession: NW side of the syncline*

North-west from the centre of the syncline the succession is repeated in the reverse order without marked variation as far as Limestone VI. At a point about two-thirds of the way from HWM to LWM, the thick sandstone between Limestones IV and V has been involved in a peculiar type of deformation. In this, the lower part of the sandstone has been forced upwards through the upper part. Strongly curved faults appear to have facilitated this, but the cause remains obscure. Laterally the disturbance dies out rapidly and the outcrop of Limestones IV and V are then unaffected.

3. *Cambo Burn to Cambo Ness; faulting*

Beyond Limestone VI the succession is probably continued down to about Limestone X or XI. It is poorly exposed but, so far as can be determined, is appreciably thinner than on the eastern side of the syncline. Limestone VII can be detected among boulders 45m east of the mouth of the Cambo Burn, but thereafter exposure is poor for the next 90m. A thin hematite-stained limestone, thought to be Limestone VIII, outcrops at HWM on the south-east side of Cambo Ness and by following this limestone down the beach, it soon becomes apparent that the area is heavily faulted. The south-east dip of 25°, seen in the higher beds of the syncline, increases to 60° and locally the beds are even overturned. A 1m shelly limestone may be examined in a small faulted patch here, but its relationship to the other limestones is unknown.

4. *Cambo Ness: sandstones, cornstones, the Humlie*

The structure at Cambo Ness is very obscure, much of the

area being overlain by large angular blocks of buff sandstone clearly not transported any great distance. Interpretation is rendered even more difficult by the fact that near LWM exposures are absent altogether. By walking a short distance to the west and north-west it will become obvious that there must be a considerable structural disturbance between Area I and Area II. This is attributed to the Cambo Fault. On the northern side of Cambo Ness the rocks are quite different to those of Area I, comprising ill-bedded, buff, reddish, greenish and grey sandstones dipping gently to the south and south-east. These carry cornstones and are sometimes cherty and are assigned to the Balcomie Beds of the Inverclyde Group.

The sandstones further to the north are cross-bedded and micaceous and are often muddy, the muddy bands being purple and green. Good exposures of these strata may be examined for 180m north-west of Cambo Ness, but thereafter are restricted to near LWM. Note here a particularly large erratic block: the Humlie. This is composed of Dalradian greenschist and measures 3 × 2 × 2m. Scattered outcrops of sandstone occur across Cambo Sands and become more numerous towards the north-western end of the sands.

5. Fault with chert; andesite conglomerate

The rocks of Area II are terminated abruptly to the north against an E–W fault, the Kingsbarns Fault. As exposed about halfway down the beach, these rocks consist of massive sandstone and knobbly weathering, massive and brecciated cornstone with veins of black chert occurring next to the fault. Near LWM, however, they are composed of sandstone and purple mudstone dipping east. These give way near the fault to a dome with marginal purple and red mudstones and, near the centre, a red conglomerate with muddy matrix. The pebbles in the conglomerate are up to 2.5cm long and consist of much-weathered andesite. This conglomerate with andesite pebbles is unusual. It implies erosion of what must almost certainly be Lower Old Red Sandstone lavas at the time the conglomerate was being laid down. The small size of the pebbles indicates considerable transport of the fragments, e.g. from the present Lower Old Red Sandstone lava outcrop in North Fife.

6. The Lecks and Kingsbarns Harbour: shelly limestone and sandstone with plant fragments

Within Area III, which extends north-westwards from the Kingsbarns Fault to another fault beyond Airbow Point, 1.5km to the north-west, the structure is very simple, the beds having a gentle inclination to the south-west and south except for a shallow basin at the Lecks near Kingsbarns Harbour. The maximum dip is 16° just north of the harbour. In the basin note the occurrence of a *Naiadites* shell bank, or biostromal limestone, 1.4m thick and outcropping as a flat bench. Possible correlations of this limestone with other Strathclyde Group limestones are unclear. Beneath the limestone the sediments are not unlike those of Area I in that, generally, they consist of alternating thick sandstones and shales. At the northern wall of the harbour, examine a sandstone which displays ripple marking, cross bedding, calcareous concretions and hematite staining. The underlying shales, eroded out lower on the beach, can be seen at HWM where ironstone nodules in the shale contain plant fragments and some of the higher shale is bituminous. They are unconformably overlain by very shelly raised beach sediment.

7. Spirorbis limestone and washouts

Forty-five metres beyond the north quay at Kingsbarns Harbour, just in front of a pipe, a 1m limestone, containing *Naiadites* and many gastropods, caps a thick bed of sandstone. Three hundred and twenty metres north-west of Kingsbarns Harbour a *Spirorbis* limestone crops out. On its wave-washed upper surface this limestone may be seen to be packed with calcareous tubes of the worm *Spirorbis*.

Beneath it a thick bed of sandstone forms a prominent ridge on the shore running for 0.75km to the north-west. Near the point where the *Spirorbis* limestone reaches HWM, two washout channels cut this sandstone roughly at right angles to the strike and break the scarp line. The more northerly of these is of particular interest, being a double washout. A later channel has removed part of the earlier one as well as part of the underlying sandstone. For the next 550m north-west, the shore runs almost parallel to the strike and only gradually do older beds appear. These comprise shales or clays with a 30cm bed of limestone 3m beneath the sandstone. This limestone is shelly at first, becoming less so to the north-west where it takes on a rubbly appearance and develops an irregular top.

8. Facies variation

Beneath this limestone, the exposures provide a very clear illustration of the variability of the strata in this type of succession. Within 450m of the double washout, a bed of sandstone makes its appearance and increases to 2m in thickness along the strike. At Airbow Point, the shales again yield *Naiadites* together with ostracods. Cornstones, some reddish in colour, occur in the sandstones which also have a tendency to take on a reddish colour. The section terminates against another fault.

The strata in Area III pass upwards into rocks strongly resembling the succession in Area I with the Randerston Limestones. As such it may be suggested that they lie beneath the Area I rocks within the Anstruther Beds, e.g. Forsyth and Chisholm (1977, p32).

Walk back to Kingsbarns Harbour and rejoin the bus. Follow the A917 back to St Andrews.

References

BROWNE, M. A. E., 1980. Stratigraphy of the lower Calciferous Sandstone Measures in Fife. *Scot. Jour. Geol.* **16**, 321–8.

FORSYTH, I. H. and CHISHOLM, J. I., 1968. The Geological Survey boreholes in East Fife, 1963–4. *Bull. Geol. Surv. Gt. Br.* No 28, 121–35.

————, 1977. The geology of East Fife. *Mem. Geol. Surv. Gt. Br.*

GEIKIE, A., 1902. The Geology of Eastern Fife. *Mem. Geol. Surv. Scot.*

KIRKBY, J. W., 1901. On Lower Carboniferous strata and fossils at Randerston, near Crail, Fife. *Trans. Edinb. Geol. Soc.* **8**, 61–75.

NEVES, R. *et al.*, 1973. Palynological correlations within the Lower Carboniferous of Scotland and Northern England. *Trans. Roy. Soc. Edinb.*, **69**, 23–70.

Pittenweem–
St Monans (half day)

OS 1:50,000 Sheet 59
GS One-inch Sheet 41
Excursion map 17.

WALKING DISTANCE: 2.4km of path, 0.8km of rocky beach.

PURPOSE: To examine Carboniferous rocks belonging to the upper part of the Strathclyde Group, in particular the Pathhead Beds (see Table IV), and the greater part of the Lower Limestone Formation, which are exposed in the St Monans Syncline. Sedimentary structures and facies variation are well displayed and are described in some detail. Simple folding and faulting can also be seen to advantage.

NOTE: It is advisable to carry out this excursion at moderately low tide, especially in the vicinity of St Monans Harbour though much of the section is exposed at half tide and there is easy access by footpath to all but the last 275m.

ROUTE: Travel south from St Andrews along B9131 to Anstruther, across rolling country with few exposures and mainly covered in glacial till. As the road descends gently to Anstruther there is a fine view across the Firth of Forth to the phonolite plugs of North Berwick Law and the Bass Rock. The Isle of May teschenite sill can also be seen 6km offshore to the south-east.

In Anstruther turn west along A917 through Pittenweem

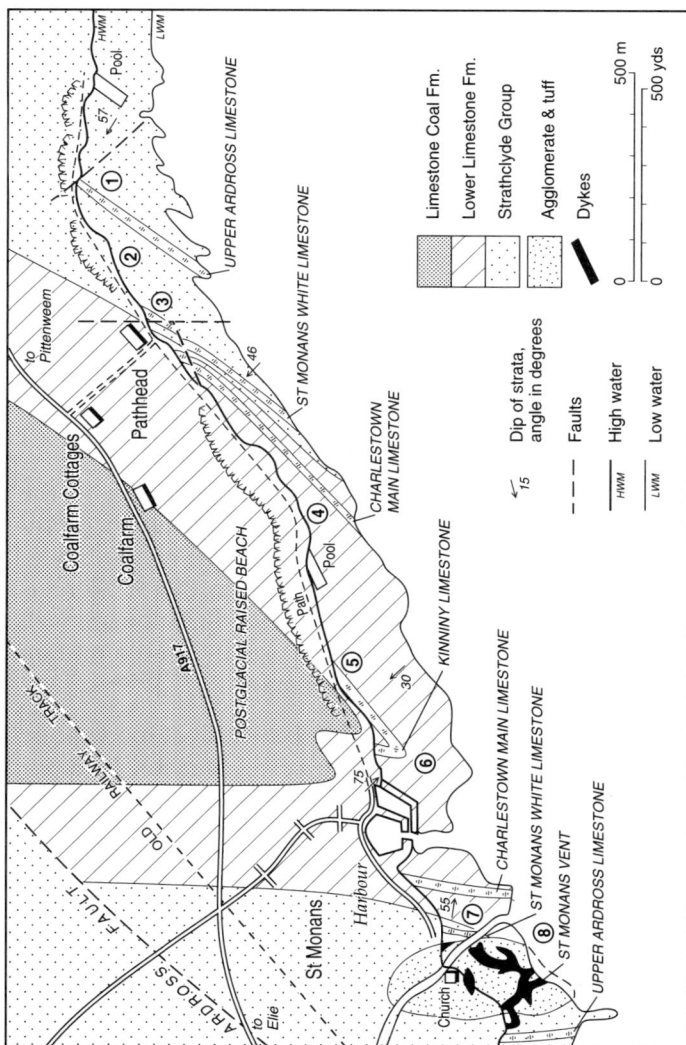

MAP 17: Pittenweem - St Monans

Legend:
- Limestone Coal Fm.
- Lower Limestone Fm.
- Strathclyde Group
- Agglomerate & tuff
- Dykes

500 m
500 yds

↙ 15 Dip of strata, angle in degrees

‒ ‒ ‒ Faults

HWM High water

LWM Low water

Map labels:
- UPPER ARDROSS LIMESTONE
- HWM
- LWM
- Pool
- 57
- ① ② ③ ④ ⑤ ⑥ ⑦ ⑧
- ST MONANS WHITE LIMESTONE
- to Pittenweem
- Pathhead
- 46
- CHARLESTOWN MAIN LIMESTONE
- Coalfarm Cottages
- Coalfarm
- POSTGLACIAL RAISED BEACH
- Pool
- Path
- KINNINY LIMESTONE
- A917
- 30
- 75
- RAILWAY TRACK
- OLD
- FAULT
- ARDROSS
- CHARLESTOWN MAIN LIMESTONE
- ST MONANS WHITE LIMESTONE
- ST MONANS VENT
- UPPER ARDROSS LIMESTONE
- Harbour
- St Monans
- 55
- Church
- to Elie

and continue as far as Coalfarm Cottages (535023). Dismount and send the bus on to St Monans Harbour, then walk down the narrow road to Pathhead (200m) which stands on a late-glacial raised beach. Follow the path down the steps to a lower postglacial raised beach and there walk 0.5km north-east towards Pittenweem to the first concrete sea wall beyond a prominent bluff of sandstone.

PART 1. THE STRATHCLYDE GROUP

The topmost 350m of the Strathclyde Group (the Pathhead Beds) are examined in this part of the excursion. The sediments are cyclical, sandstones making up 60–70 per cent of the succession. Lesser amounts of shales and sandy shales are present with coal and calcareous beds accounting for a very small percentage. With the exception of parts of the St Monans White Limestone and the St Monans Brecciated Limestone (of the Lower Limestone Formation), all the limestones in the section have been dolomitised. The succession is tabulated below:

	Metres
Dark grey shale	5.0
Sandstone and seatearth heavily bioturbated	3.0
Shales becoming more sandy upwards with fossil-iferous band (containing *Camarotoechia*) 2m from base; many ironstone bands	7.3
ST MONANS WHITE LIMESTONE with *Lithostrotion* and brachiopods abundant; buff weathering 0.7m dolomitic band in centre; crystalline dolomitic band making up the lowest 0.7m.	4.3
Shales with bivalves	1.2
Sandstone, thinly bedded in places	0.9
Sandstones and shales; bioturbated in top 2m	2.5
Thin coal with seatearth	0.7
Interbedded sandstones and shales with ironstone and calcareous bands	6.5
Sandstones with seatearth towards base and some bioturbation towards top	7.3

PATHHEAD UPPER MARINE BAND; contorted
shaly and sandy beds with *Lingula*   6.0

Sandstones with roots and thin bedding   3.6

PATHHEAD LOWER MARINE BAND; shales with
thin limestones and carbonate nodules, corals,
brachiopods, bivalves and crinoids*   1.2

Sandstones with clay and seatearth beds and ripple
cross bedding   9.7

Yellow weathering sandstones with ripple cross
bedding and convoluted bedding   5.5

Sandstones, clays, thin coals and seatearth (not
visible)   4.6

Sandstones with cross bedding and convolute
bedding   21.3

Sandstones, clays, thin coals and seatearth;
bioturbated and with distinct washout at HWM   4.6

Yellow and grey sandstone, shaly in places with
ripple cross bedding, ironstone nodules and plant
rootlets; bottom 1m bioturbated   2.1

Grey shale with ironstone nodules   9.1

UPPER ARDROSS LIMESTONE: buff weathering
dolomite with brachiopods, crinoids and ostracods   0.6

Shale and coaly shale with coal (obscured at HWM
by loose material from drift deposit above)   0.9

Sandstone with rootlets towards the top   2.3

Shales with ironstone and cementstone bands and
nodules; abundant brachiopods, bivalves and
gastropods, sandstone bands appearing nearer top 13.1

LOWER ARDROSS LIMESTONE; buff weathering
dolomite with some minor thrusting   0.5

Sandy grey shale with abundant *Lingula*   3.7

Shale with plant fragments and slickensides   0.9

Oil shale with yellow efflorescence (forms very hard
ridge lower on shore though with no efflorescence) 0.5

Interbedded sandstone and shale with some trough
cross bedding; sandstone finely bedded with ripple
cross bedding   0.6

Pale grey shale with plant fragments and small
carbonate nodules   0.9

Coal   0.15

Bleached white sandy seatearth with rootlets and
plant remains   0.9
Hard yellow sandstone   0.3
Sandstone and shale with ripple cross bedding and
bioturbation   0.3
Hard, yellow, cross bedded sandstone   1.5
Sea wall – no further exposures

* These are known locally as 'Croupies' and, in days gone by, were taken to sea by fishermen as a defence against drowning.

1. The Ardross Limestones

The sequence from the sea wall to the Upper Ardross Limestone is well exposed and should be examined carefully. The table shows that the sediments in the lower part of the two cycles differ, particularly in the thickness of the beds that lie between each coal and the limestone above it. In the first case, 6.6m of varied strata are present between the coal and the

FIGURE 11: The well developed sedimentary 'cycle' at the horizon of the Lower Ardross Limestone between Pittenweem and St Monans. Sandstone at the base of the cycle forms the ridge on the right, the Ardross Limestone the ridge near the centre and the sandstone at the base of the next cycle is on the left.

Lower Ardross Limestone whereas only 0.2m of shale occurs between the coal and the Upper Ardross Limestone (although obscured by detritus, the position of the coaly horizons may still be picked out). It is believed that the difference reflects different rates of subsidence at the ends of the two periods of coal formation: slow in the first case and more rapid in the second (Greensmith 1965, p234). An examination of the shales beneath the Lower Ardross Limestone points to the intervention of lacustrine conditions (oil shale) followed by more marine conditions (*Lingula*-bearing shales) between the forest-swamp coal and the clear-water marine limestone.

2. Thick sandstones with convolute bedding; soft sediment slumping

Between 45m and 90m west of the Upper Ardross Limestone, at the prominent sandstone bluff mentioned earlier, it will be seen that a 21m-thick sandstone rests on a ripple and cross-bedded sandstone in which carbonaceous fragments accentuate the bedding. In the 21m-thick sandstone both the bedding and the cross bedding give way upwards to convolute bedding on a large scale. This persists to within 2m of the top of the sandstone before there is a return to normally bedded sandstone. The convolution of the bedding is believed to be due to fluidisation of the unconsolidated sand shortly after deposition. At the present day this sandstone has been planed off by wave action to produce a remarkably level wave-cut platform.

Ninety metres further west at a small grassy point on the shore, examine a belt of disturbed rocks which extends along the strike for 150m from HWM. Above and below it the sandstones dip steadily north-west at between 40° and 60° whereas within the disturbed area, large sandstone masses have been rotated dextrally to lie inclined to the general strike. The interbedded shales have deformed plastically around them. This deformation is most probably caused by slumping in the unconsolidated sediments, as has been described in modern sediments on the Mississippi delta.

3a. Pathhead St Monans White Limestone

At HWM on the beach below Pathhead Cottage the St Monans White Limestone is well exposed and is conspicuous on account of its white colour. The beds below it also merit examination. Four metres below the limestone the sandstones are intensely bioturbated with *Teichichnus* and are overlain by a metre of fossiliferous grey shale with, at the top, a 5cm thick oil shale showing at HWM the characteristic yellow efflorescence due to the mineral jarosite, an iron-bearing sulphate.

The ledge-forming, lower parts of the St Monans White Limestone are brownish, crinoidal and dolomitic, while the higher, white parts are a richly fossiliferous *Lithostrotion*-bearing coral limestone with, less commonly, the brachiopods *Athyris*, *Schizophoria*, *Camarotoechia* and '*Spirifer*'. There is a conspicuous 0.8m thick brown, dolomitic bed running through the white outcrop – also with *Lithostrotion*. Some 10m of mainly marine shales overlie the limestone.

PART 2. THE LOWER LIMESTONE FORMATION

At Pathhead the base of the Lower Limestone Formation is reached and the succession within that group should be considered before examining local details. The succession within the St Monans Syncline is tabulated below (thicknesses after Forsyth and Chisholm 1968). Recently Fielding *et al.* (1988) have examined the seatearths in the section, concluding that lower delta plain conditions had allowed the formation of substantial coal seams from time to time.

	metres
Sandstone	1.0+
Dark grey fossiliferous marine shale with ironstone nodules	5.0
MIDDLE KINNINY LIMESTONE with *Zoophycus* markings near the top	1.5
Sandstones with coals up to 70cm thick, seatearths	

and shales up to 2m thick: abundant *Teichichnus* burrows at swimming pool	25.0
Well jointed sandstone	2.0
Fossiliferous grey shales with nodule bands: only partly exposed and forming trough on beach at St Monans swimming pool and harbour mouth (horizon of the Seafield Marine Band)	12.0
Sandstones with Largoward Black Coal at seaward wall of swimming pool	15.0
Shale including thin sandy Mill Hill Marine Band	1.5
Sandstones becoming silty downwards	12.0
Grey shale with ironstone nodules, and sandy beds increasing upwards	22.0
Neilson Shell Bed in shale	1.5
CHARLESTOWN MAIN LIMESTONE: dolomitic, grey and crinoidal with impersistent thin shale band in the middle, and bioturbation.	1.5
Shales with seatearth at base	0.9
Massive sandstone with abundant large plant fragments at top	2.3
Sandstones, mudstones, seatearths and thin coals (Radernie Coals)	4.6
Sandy, micaceous shales with ironstone nodules and thin sandstone at top	4.3
ST MONANS LITTLE LIMESTONE: crinoids and large productids, 5cm thick hematite-stained band at top	0.9
Sandy shales	1.5
Sandstones with cross bedding, ripple cross bedding and convolute bedding* ⎫ Sandy shales with sandstone laminae and ironstone nodules* ⎬ Grey shales with ironstone nodules* ⎭	11.0
Micaceous sandy shales	2.7
ST MONANS BRECCIATED LIMESTONE: nodular with thin shale laminae	3.5

* Largely cut out at HWM by the Pathhead Fault, but present near LWM and on the west side of the St Monans Syncline. Combined thickness is 11.0m

*3b. Pathhead: the Pathhead Fault; Lower Limestone Formation
facies variation*

At Pathhead the succession seen at HWM does not represent
the true thickness of the beds on account of faulting.

The Pathhead Fault runs into the cliff in three branches
which lie in the shaly parts of the succession above and below
the St Monans Brecciated Limestone and above the St Monans
Little Limestone. The main fault striking N–S has a marked
sinistral offset of some 18m on the shore and there is considerable disturbance along its length. A WSW-striking branch
from the Pathhead Fault has a dextral offset and cuts the St
Monans White Limestone, the St Monans Brecciated Limestone and the St Monans Little Limestone. The fault offsets
the last by 6m and also cuts the overlying 2.3m massive
sandstone. Westwards, however, almost all the movement is
taken up in the shales beneath the Charlestown Main Limestone, and accordingly the limestone is displaced by only a
few centimetres. Note the conspicuous gap where the fault
cuts the ridge formed by the massive sandstone.

The St Monans Brecciated Limestone, which forms the base
of the Lower Limestone Formation, is a nodular limestone
containing many seams of calcareous shale or clay which
branch and reunite round limestone nodules which may be
up to 25cm thick and 25 to 50cm long. This character is
maintained along the length of the outcrop and has resulted
in the limestone usually being eroded out to form a trough
on the beach (in contrast to the other limestones which form
ridges). The name 'Brecciated' is due to the fragmented nature
of the limestone adjacent to the Pathhead Fault. This brecciation, however, is found for only 1m from the fault and nowhere else. Forsyth and Chisholm (1977) have correlated this
limestone with the Hurlet Limestone as the basal unit of the
Lower Limestone Formation. The St Monans Little Limestone
contains giant productids which are not, however, readily
collected. The topmost 5cm are hematite stained and the remainder weathers to a pale grey/buff colour.

At HWM above the St Monans Little Limestone several

metres of seatearths mark the position of the Radernie Coals, up to 1.3m thick inland, but here only a few centimetres thick.

The Charlestown Main Limestone is split into an upper 90cm bed and a lower 60cm bed by an impersistent 3cm shale seam.

South-west from Pathhead, the Charlestown Main Limestone is succeeded by 21m of moderately fossiliferous dark grey shales with carbonate nodules. At the base of these shales Wilson (1966) has recognised the Neilson Shell Bed faunal assemblage including *Straparollus carbonarius* and *Posidonia corrugata gigantea*. A considerable thickness of sandy beds succeeds the shales. Upwards the sandstone bands increase in abundance and thickness with sedimentary structures being well displayed. These include starved ripples, an erosion surface with shale clasts and, unusually, a sandstone with flute casts. A thin limestone is followed by black pyritous shales and sandstones with concretions up to 2m across (these beds include the Mill Hill Marine Band horizon).

4. Swimming Pool: sedimentary structures

Continue along the path to the west end of the swimming pool, noting very prominent jointing in the sandstone which forms the north side of the pool. The pool itself stands on a 10m-thick marine shale band – the Seafield Marine Band horizon of the Kirkcaldy area (Chisholm 1970). At the steps at the west end of the pool, it will be seen that the sandstone is intensely bioturbated with trace fossils including *Teichichnus* and *Planolites* (Chisholm 1970).

5. Middle Kinniny Limestone; the trace fossil Zoophycus

One hundred and sixty metres west of the pool and 4m south of the path (which lies here on a low postglacial raised beach) lies the Middle Kinniny Limestone which is 1m thick and displays *Zoophycus* markings in abundance in the top 15cm. These can be seen both in section and when the rock is split along the bedding and particularly in the fallen blocks.

Observe the prominent cliff at the back of the raised beach. On top of this cliff, a terrace of houses and an old windmill stand on a higher late-glacial raised beach. The windmill, restored in 1992, formed part of the equipment for working the saltpans, the foundations of which can be seen adjacent to the coast path. Coal from mines in the vicinity was extracted from either the Limestone Coal Formation or the top of the Lower Limestone Formation for use in the saltworks. The mines were abandoned about 1800 and there is no accurate information available (Forsyth and Chisholm 1977, p86). The salt was exported from Anstruther.

6. St Monans Syncline

Follow the outcrop of the Middle Kinniny Limestone west for 320m to the axis of the St Monans syncline. The change of dip at the axis is very noticeable and takes place within about 3m, the limestone in particular being shattered at the axial trace of the fold. The change in dip is from 38° NW to 70°ESE. The line of broken rocks on the axial trace can be followed down the shore in a shallow trough eroded out by the sea. The top part of the succession is then repeated in reverse order from here to the harbour wall.

The next feature can only be seen clearly at low spring tide. This is a small anticline within the main syncline and exposed just east of the harbour mouth.

7. The Lower Limestone Formation on the west side of the St Monans Syncline

Return to the shore at the west side of the harbour. The succession is obscured by rubbish at HWM and must be examined lower on the shore. The beds are not markedly different to those on the east limb of the St Monans Syncline and the following notes may be helpful in working down through the succession.

The Charleston Main Limestone occasionally swells to as much as 2.5m while the massive sandstone beneath again forms a conspicuous ridge on the shore and still contains abundant plant remains towards the top. The St Monans Little Limestone is again characterised by the hematite-stained top.

213

Between these limestones is a small anticlinal fold composed of sandstone causing a prominent hump on the shore. The St Monans Brecciated Limestone weathers out as a gully and is thus not easily detected although it occurs at the expected horizon. The St Monans White Limestone forms a prominent ridge, particularly near HWM. It contains the same brown dolomite band, but is no longer a coral limestone as it was at Pathhead; instead it is more compact and crinoidal. The shales overlying it are particularly fossiliferous and at HWM an impersistent 25cm limestone band 1.2m above the base is packed with *Camarotoechia*, '*Productus*', '*Spirifer*', crinoid ossicles, bivalves and bryozoa. The beds beneath the St Monans White Limestone (comprising sandstones with seatearths, sandy shales and thin coals) are somewhat disturbed by faulting. Fourteen metres of these beds crop out before the St Monans Neck interrupts the succession.

8. Disturbed strata adjacent to the St Monans Neck

At low spring tides the St Monans White Limestone can be traced across the St. Monans Burn. It strikes almost due south and dips 45°E before being cut off by the neck as are the various sandy beds beneath. The St Monans Brecciated Limestone can also be traced across the burn until it too is cut off by the neck. A short distance seawards both limestones reappear dipping at about 60° NW off the flank of a sharply folded anticline which plunges steeply NNE. The strike of the sediments swings round in a few metres to almost north–south, dipping steeply east. An ENE–WSW trending fault throws these beds about 30m east where they reappear as a series of reefs lying offshore. From aerial photographs it can be seen that the strata maintain their north-south strike past the St Monans Neck on its seaward side and are continuous with the rocks on the western side.

Walk back to St Monans Harbour to rejoin the bus and return to St Andrews.

References

CHISHOLM, J. I., 1970. *Teichichnus* and related fossils in the Lower Carboniferous of St. Monance, Scotland. *Bull. Geol. Surv. Gt. Br.* No 32, 21–51.

FIELDING, C. R. *et al.*, 1988. Deltaic sedimentation in an unstable tectonic environment – the Lower Limestone Group (Lower Carboniferous) of East Fife, Scotland. *Geol. Mag.*, **125**, 241–55.

FORSYTH, I. H. and CHISHOLM, J. I., 1968. Geological Survey boreholes in the Carboniferous of East Fife, 1963–4. *Bull. Geol. Surv. Gt. Br.* No 28, 61–101.

————, 1977. The geology of East Fife. *Mem. Geol. Surv. Gt. Br.*

GREENSMITH, J. T., 1965. Calciferous Sandstone Series sedimentation at the eastern end of the Midland Valley of Scotland. *Jour. Sed. Petr.* **35**, 223–42.

WILSON, R. B., 1966. A study of the Neilson Shell Bed, a Scottish Lower Carboniferous marine shale. *Bull. Geol. Surv. Gt. Br.* No 24, 105–30.

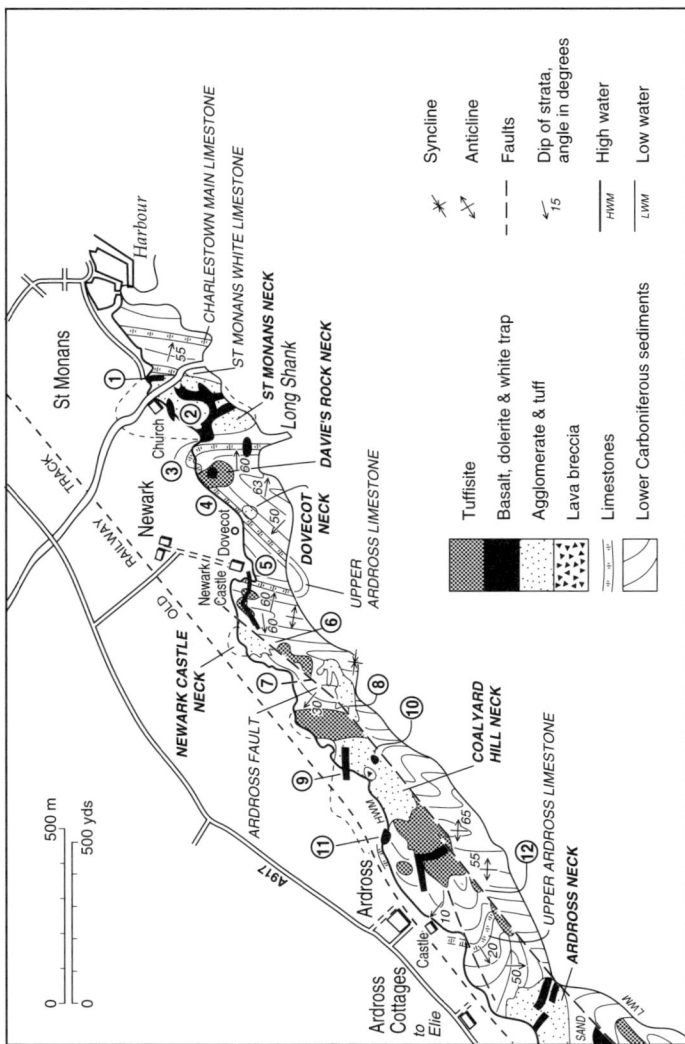

MAP 18: St Monans - Ardross.

Legend:

- Tuffisite
- Basalt, dolerite & white trap
- Agglomerate & tuff
- Lava breccia
- Limestones
- Lower Carboniferous sediments

Symbols:

- Syncline
- Anticline
- Faults
- Dip of strata, angle in degrees — 15
- High water — HWM
- Low water — LWM

Labels on map:

St Monans
Harbour
CHARLESTOWN MAIN LIMESTONE
ST MONANS WHITE LIMESTONE
ST MONANS NECK
Long Shank
DAVIE'S ROCK NECK
Church
Newark
DOVECOT NECK
UPPER ARDROSS LIMESTONE
TRACK
RAILWAY
Newark Castle
Dovecot
NEWARK CASTLE NECK
ARDROSS FAULT
COALYARD HILL NECK
UPPER ARDROSS LIMESTONE
ARDROSS NECK
Ardross
Ardross Cottages
to Elie
Castle
SAND
A917
500 m
500 yds

216

St Monans–Ardross (half day)

OS 1:50,000 Sheet 59
GS One-inch, Sheet 41
Excursion map 18.

WALKING DISTANCE: 2km of rocky beach.

PURPOSE: The main objects of this excursion are to examine: (1) Carboniferous sediments belonging to the highest part of the Strathclyde Group, the Pathhead Beds; (2) a series of Carboniferous volcanic necks and their relationship to the country rocks; (3) the relationship between the Ardross Fault and the volcanic necks.

ROUTE: As for Excursion 12, but proceeding direct to St Monans Harbour and thereafter walking along the shore section to rejoin the bus which should be sent on to the lay-by at Ardross Cottages (506006) 1.5km south-west of St Monans. This excursion is seen to best advantage at low tide since many of the exposures are on the wavecut platform.

1. St Monans Neck: eastern margin

Follow the road west then south-west for 180m from the harbour before descending to the shore at a rough cart track cut in the rocks 40m short of the St Monans Burn. Sandstones lying about 15m below the St Monans White Limestone outcrop here and dip 45° SE. Now continue along the track towards St Monans church. Adjacent to the St Monans Neck it will be seen that the sandstones are baked and shattered and that the dip steepens abruptly at the edge of the neck. On crossing the neck margin, at the top of the beach notice a 1.2m basalt dyke immediately inside the neck. It extends only a few metres downshore. Beyond it the exposures consist of coarse tuff or agglomerate containing basalt blocks up to 30cm across. Westwards this tuff becomes less coarse as it is traced into the neck. At the western end of the churchyard wall notice a 30cm thick dyke which pinches out and is then continued *en echelon* 30cm away. Downshore the eastern margin of the neck makes a pronounced feature: a wavecut platform of tuff standing a few metres above the level of the St Monans Burn which here crosses poorly exposed, folded sediments. The even surface of this platform is broken by protrusions caused by resistant calcite veins, small basic dykes and large basalt blocks within the tuff. There is little sign, however, of bedding in the tuff.

2. St Monans Neck: vent intrusion

Seventy metres SSE of St Monans Church notice a prominent trench which crosses the wavecut platform. This with its minor branches marks the course of a large basalt vent intrusion or dyke. It has been preferentially eroded by the sea because of its well developed columnar jointing normal to the dyke walls. The form of the intrusion at the seaward side of the neck is obscure, but several stacks composed mainly of tuff, perhaps baked by the dyke, project as much as 6m above the wavecut platform there. The intrusion can then be traced north-west as a dyke to the neck margin and beyond where it cuts at right angles through the 'Long Shank', a prominent sandstone ridge. The dyke has been preferentially eroded here too, but some of its chilled margin can still be seen adhering to the sandstone.

3. St Monans Neck: western margin

Outside the neck the dyke bends round to the north and becomes concordant with the sediments which here dip

FIGURE 12: The St Monans Neck seen across the St Monans Burn. The level wave-cut platform is underlain by tuff, while the trench marks the site of a dyke-like vent intrusion. In the background are sea stacks of tuff, dark, and sandstone beyond the neck, light in colour.

steeply to the east. It has two branches both altered to white trap: one in the sandstone and the other in a coaly bed at the junction between the sandstone and the underlying shales.

Along the north-western part of the neck margin the intrusive nature of the tuff can be clearly seen, an intimate mixture of tuff and sandstone being exposed in the low cliff at the top of the beach. In the same low cliff 50m to the east is a small dyke that dies out downwards. The tuff, being free of seaweed, is also best examined around here. The material in it is mainly of igneous origin: large fragments of basalt lying in a matrix of smaller fragments of similar material grading down to dust size. Pieces of sediment are relatively rare, but pieces of pale weathering limestone and also coaly masses veined with calcite (which were originally wood) do occur.

Study of aerial photographs indicates that the southern margin of the neck must lie just south of LWM. The northern margin is not exposed, but tuff extends at least 45m north of the church. The neck is entirely filled with tuff or agglomerate except at its western margin where several large masses of sediment occur within it.

4. *Ardross Limestones; Davie's Rock Neck; the Dovecot Neck*

The next bay to the west is occupied by a NE-plunging anticline with dips of 60°+E and 40°–60° NW. The Lower and Upper Ardross Limestones, both of which are crinoidal, can be seen on both limbs, together with a 30cm-thick oil shale (displaying a yellow efflorescence), which occurs beneath the Lower Ardross Limestone as at Pathhead east of St Monans (Excursion 12). Well displayed in the cliff at HWM is the Upper Ardross Limestone, which has been crumpled and broken over the crest of the anticline. This limestone, which consists of two distinct beds separated by 15cm of shale, lies in a thick shale sequence. The shales are richly fossiliferous, especially in the vicinity of the limestone (which also contains many fossils although these are difficult to extract). The core of the anticline may be examined low on the shore where several prominent sandstone ridges are tightly folded, sheared and crumpled.

Two volcanic necks cut the anticline: an eastern one, Davie's Rock Neck, and a better known one further west, the Dovecot Neck. Davie's Rock Neck contains a plug of 'basalt' which forms the stack after which the neck is named. A careful examination of the contact here will show that the basalt has been altered to white trap adjacent to the surrounding shales while these have been brecciated for a few metres outwards from the basalt. The greater part of the neck, however, is occupied by broken up sediments or tuffisite through which run small and continuous white trap dykes. Several similar dykes which occur in the surrounding shales are probably associated with the neck though they can seldom be traced directly into it. One of these can be seen in the cliff just north of the basalt plug; another in a minor fault a few metres west of the neck; and two more to the east striking E–W across the Ardross Limestones.

The Dovecot Neck (519012) is considerably smaller than the Davie's Rock Neck, being only 55m long and 36m wide, but it too forms a feature on the shore. It lies at HWM and contains a group of enormous blocks of sandstone including one measuring 22m × 6.5m. Along with smaller pieces of sandstone and shale, together with ironstone nodules, these blocks are set in a matrix of fine-grained, pale green tuff. The eastern margin of the neck is sharply defined and cross cutting whereas the western margin wedges along bedding planes into the country rock. A subsidiary pipe-like mass or offshoot of tuff lies to the south-east of the main neck. Two dykes cut the neck, one of which continues to the south-east for 90m across the anticline. Minor white trap dykes also occur on the north-west of the neck and the adjacent sandstone is cut by veins of pyrite and barytes. The neck cuts both the Upper and the Lower Ardross Limestones.

5. Newark Castle: deltaic sandstone with slumping; tuffisite

Newark Castle (518012) stands on a synclinally folded, thick sandstone in which cross bedding and convolute bedding are well displayed (this sandstone is also exposed in the Long Shank and between Pittenweem and St Monans – Excursion 12, Location 2). It will be seen from the map that dips are steep in this fold and that to the west the Ardross Limestones reappear from beneath the sandstone. West of the castle, in the shales between the Upper and Lower Ardross Limestones, a white trap dyke cuts across the strike, runs through the Upper Ardross Limestone and dies out in the sandstone cliff to the east. Two sills of tuffisite, consisting of sedimentary fragments in a hard shaly matrix, occur a short distance below the Upper Ardross Limestone. Notice that the higher of these has been intruded mainly beneath the Lower Ardross Limestone, but that it transgresses this limestone at one point to lie in part above it as well. The white trap dyke mentioned above cuts the two tuffisite sills and can be followed to the west where it also runs through a small volcanic neck including a basalt plug within which the dyke rock is unaltered.

The dyke finally dies out in the sediments beyond. Folding is well displayed in the sediments a short distance downshore from this small neck.

6. *Ardross Fault: Newark Castle Neck; dykes*

The Ardross Fault is first encountered at the grassy point 130m west of Newark Castle. From the map it can be seen that the sediments on the south-east side of the fault are intensely folded while those on the north-west side are not. Francis and Hopgood (1970, p181) have suggested that this is due to a fairly considerable downthrow on the north-west side bringing relatively high level structures opposite relatively low level structures.

A series of volcanic necks is aligned along the fault and it will be noticed that most of these appear to be truncated by it although at distances of 90m and 140m south-west of the Coalyard Hill Neck, two very small necks appear to lie in or

FIGURE 13: The Ardross Fault, running directly away from the observer, separates dark tuffs within the Coalyard Hill Neck on the left from steeply dipping Carboniferous sediments on the right. In the background are the ruins of Newark Castle.

very close to the plane of the fault. Cumming (1936, p351) put forward the view that the Ardross Fault cuts the necks, displacing them by 1200m dextrally, and is therefore younger than the necks. More recently Francis and Hopgood (1970, pp179–84), while confirming the dextral component and timing of the fault relative to the necks, did not accept his amount of lateral movement and suggested that a strong vertical component was also involved. Upton (1982, p268) agreed broadly with Francis and Hopgood and suggested that the necks utilised a zone of weakness which can also be linked to the fault zone. The remainder of the excursion is designed to examine: (1) the phenomena within and adjacent to the volcanic necks on this part of the shore; (2) some of the features displayed on the shore which have given rise to the above views.

The Newark Castle Neck is filled with dull green tuff and its boundaries are difficult to delineate except on the south-eastern side where it ends against the Ardross Fault, shales being exposed beyond the fault plane. Francis and Hopgood (1970, p177) suggested that here the Ardross Fault is coincident with the original neck margin. The course of the Ardross Fault is visible on this part of the shore on account of differential erosion of the sandstone and tuff, or sandstone and shale, which are in juxtaposition across the fault. On the south-eastern side of the fault, beginning 36m south-west of Newark Castle Neck and opposite a very conspicuous vertical sandstone ridge, is a 115m long irregular neck filled with intrusion breccia. Just outside the western end of this neck, a white trap dyke ends abruptly against the fault thus providing time relations of the dyke emplacement and faulting. Now walk north-east back along the fault towards the sandstone ridge where the margin of the small neck will be seen to end abruptly against the fault. Some 45m east from the above white trap dyke much white trap is exposed apparently lying in the fault plane. A close examination of this suggests, however, that it too may have been sheared by the fault.

7. Coalyard Hill Neck: faulting in country rock; eastern margin of neck

At this locality the cliff at HWM provides good examples of faulting in gently dipping sandstone. From this point walk west until the north-eastern margin of the Coalyard Hill Neck is reached. The edge of the neck forms a distinct step on the shore platform, tuffisite within the neck standing above the level of the shales in the country rock to the east. From the map it will be seen that there is a large area of tuffisite at both the south-western and north-eastern ends of the neck. When the neck margin is examined in detail, it will be seen that the tuffisite ramifies into the country rock and that it is locally impossible to draw a distinct boundary between them. Similar phenomena can be seen as the neck margin is followed south to the Ardross Fault. Forsyth and Chisholm (1977, p196) also regard the Ardross Fault here as being coincident with the original margin of the Coalyard Hill Neck over a distance of 100m.

8. Tuffs at the Ardross Fault

At this locality tuff is present on both sides of the Ardross Fault. To the south-east the tuff, which extends eastwards for 180m to a basaltic plug, shows a marked alignment as do large sandstone xenoliths and Francis and Hopgood (1970, p176) have pointed out that this appears to have been controlled by the strike of the country rocks which have been invaded and replaced. The tuff of the Coalyard Hill Neck on the north-western side of the Ardross Fault, while again having a major sediment component, lacks the strong alignment of the south-eastern side. Francis and Hopgood (1970, p175) have concluded that this tuff, which they refer to as part of the outer neck of the Coalyard Hill Neck, has been little affected by faulting (1970, p181).

9. Coalyard Hill Neck; dyke and tuff

Walk west across the Coalyard Hill Neck to a point at HWM (513009) where a NNW trending basalt dyke forms a conspicuous feature on the shore. The dyke has been emplaced in the tuff and has, in turn, been cut by later veins of tuff. Xenocrysts of anorthoclase and hornblende occur in the dyke rock. In this part of the neck the tuff comprises mainly fragments of basalt, but scattered dove-grey limestone fragments are present and in part the tuff has a distinctly grey colour due to the content of pulverised shale. In the main, however, the tuff is of igneous origin and it forms part of Francis and Hopgood's (1970, p175) inner neck in the Coalyard Hill Neck. Irregular calcite veins are widespread.

Ninety metres SSW, in the first stack, notice the coarse agglomerate within the neck. In it blocks of basalt up to 30cm across are particularly abundant and are not infrequently rounded. The rock may be described as a lava breccia.

10. Coalyard Hill Neck: Ardross Fault; 'basalt-capped' stack

From the lava breccia walk east for 40m to the second stack – the 'basalt capped' stack (513008) – where the bulk of the stack comprises tuff and the cap basalt. The seaward margin of the Coalyard Hill Neck is particularly well displayed in this area where it appears as a line running along the Ardross Fault, accentuated by differential marine erosion of the sediments and tuff on opposite sides of the fault. The sediments to the south-east show signs of drag suggesting dextral movement on the fault. Starting 65m south-west of the 'basalt capped' stack, the course of the Ardross Fault becomes difficult to follow. Also in this vicinity and within a few metres of the neck margin, the tuff displays a rough alignment of the fragments. Francis and Hopgood (1970, p173) suggest that the tuffs are aligned parallel to the fault owing to their being in a shear zone. A careful examination of the fault-bounded margin of the Coalyard Hill Neck, 275m south-west of the 'basalt capped' stack, shows two *en echelon* faults, a north-east one dying out to the south-west in the sediments and a south-west one dying out to the north-east in the tuff.

225

11. Coalyard Hill Neck margin

Returning once more to the top of the beach, the margin of the Coalyard Hill Neck can be examined 230m east of Ardross farmhouse at HWM (510008). Notice that the country rock has been disrupted. A small basalt plug (6m across) is exposed just inside the neck and is packed with lighter coloured xenoliths of peridotite, carbonated peridotite and, to a lesser extent, pyroxenite. Both they and the basalt are cut by numerous carbonate veins. Seventy-five metres east of here and 15m south of HWM may be found another small basalt mass containing scattered, large, yellow anorthoclase xenocrysts. It has been cut by several late tuffisite veins. The tuff in which this mass is emplaced is similar to that seen at Locality 9 with basalt blocks, usually less than 8cm across, set in a tuffaceous matrix. Where the neck margin reaches HWM, in an upstanding mass, the tuff contains several large blocks of recrystallised *Lithostrotion* limestone. Various authors, including Cumming (1936) and Forsyth and Chisholm (1977), believe these to be pieces of the St Monans White Limestone in which case they must have fallen some 60m down the neck since immediately outside the neck at this point are exposures of the Lower Ardross Limestone. This is known elsewhere to lie about that distance below the St Monans White Limestone. The 'Shrimp Band' lies 3.5m below the Lower Ardross Limestone and dips gently north. It is difficult to locate and collecting from it is also difficult. For 275m west the exposures consist of shale cut by a small tuffisite neck, by minor faults and by several tuffisite dykes a few centimetres wide.

To the south is a large area of tuffisite forming the southwestern end of the Coalyard Hill Neck. The surface of this too stands a metre above the adjacent shales. Observe that the bedding of the country rock can often be traced for considerable distances into the tuffisite. This part of the Coalyard Hill Neck also ends abruptly against the line of the Ardross Fault.

12. Small necks on the line of the Ardross Fault

Now follow the Ardross Fault south-west beyond the end of

the Coalyard Hill Neck. The first of the minor necks is reached 90m to the south-west. It displays the following features: (1) the sandstones on the south-eastern side of the fault are strongly sheared; (2) the tuffs within the neck, though possibly sheared, appear to be less so than the country rock; (3) the white trap crosses the fault plane with no sign of being disturbed. Francis and Hopgood (1970, p175) suggest that almost all the volcanics predate the fault (hence are sheared). This white trap dyke, however, gives the appearance of being later than the fault.

A few metres further to the south-west another tuffisite neck measuring 90 × 18m is bounded by the fault at its north-eastern end, but diverges from it when traced to the south-west. From this neck onwards, footing is difficult low on the shore. Follow the outcrop of the Upper Ardross Limestone up the shore. Adjacent to the neck it is folded into a tight anticline and syncline, possibly formed at the time of the main fault movements. Further up the beach there is a gentle south-west dip which continues to the cliff at HWM. There the Upper Ardross Limestone and the overlying shales are richly fossiliferous and have yielded the brachiopods *'Productus'*, *Athyris*, *Schizophoria* and *Lingula*, bivalves and many gastropods including *Euphemites* and *Soleniscus*. Next examine the cliff exposures to the east. A series of sandstones underlies the limestone. Two thick beds of sandstone are separated by a thin coal seam with seatearth and the top of the lower sandstone displays well developed ripple marking. Exposed from time to time at the base of the cliff is a thin bed of sandstone with interesting slump structures.

Either continue south-west onto Excursion 14 or walk 200m south-west to the track up to Ardross Cottages to rejoin the bus for the return journey to St Andrews.

References

CUMMING, G. A., 1936. The structure and volcanic geology of the Elie–St Monans district, Fife. *Trans. Edinb. Geol. Soc.*, **13**, 340–65.

FORSYTH, I. H. and CHISHOLM, J. I., 1977. The geology of East Fife. *Mem. Geol. Surv. Gt. Br.*

FRANCIS, E. H. and HOPGOOD, A. M., 1970. Volcanism and the Ardross Fault. *Scot. Jour. Geol.* **6**, 162–85.

UPTON, B. G. J., 1982. Carboniferous volcanism. In Sutherland, D. (Ed.) *Igneous rocks of the British Isles.* 255–75. Wiley, London.

Excursion 14

Ardross–Elie Harbour (half day)

OS 1:50,000, sheet 59
GS One-inch, sheet 41
Excursion map 19.

WALKING DISTANCE: 2.5km of rocky and sandy beach.

PURPOSE: The main objects of this excursion are: (1) the examination of a series of Carboniferous volcanic necks exposed along the shore; (2) to examine Lower Carboniferous sediments belonging to the highest Strathclyde Group (Pathhead Beds) and the Lower Limestone Formation; (3) to study the south-western end of the Ardross Fault.

NOTE: The excursion follows on from Excursion 13 and may be combined with it, but it is advisable to visit this section during low tide.

ROUTE: As for Excursion 12, but continue on the A917 past St Monans to the lay-by at Ardross Cottages (506006). Follow the path to the beach and send the bus on to Elie harbour.

1. Ardross Cottages: faulting

The rocks exposed on the shore here are sandstones within the Pathhead Beds and lying about 25m above the Upper Ardross Limestone which crops out at HWM 150m to the east (see Excursion 13). These sandstones display cross bedding and slumping and are cut off downshore a short distance

229

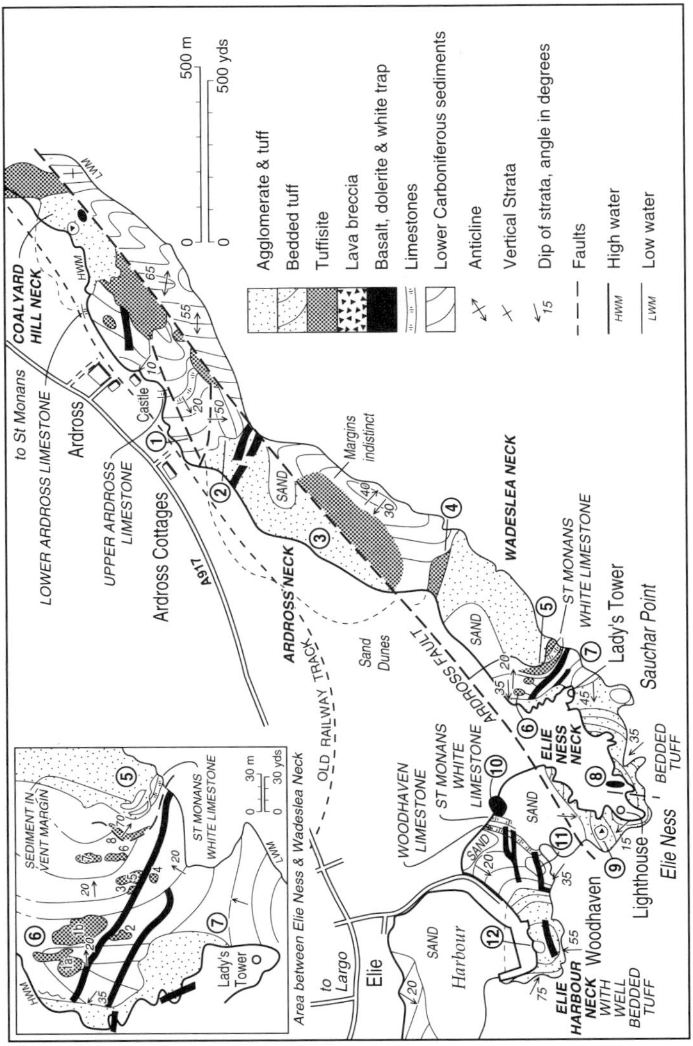

MAP 19: Ardross - Elie Harbour.

south of the bridge by a wrench fault branching off from the Ardross Fault. There is a marked topographic break along the line of this fault: a much more shaly sequence lying on the south side. On this part of the shore, the Ardross Fault itself can only be seen at extreme low spring tide. A mass of crinoidal limestone on the line of the fault was believed by Cumming (1936, p343) to be one of the Ardross Limestones.

2. Ardross Neck margin and dykes in neck

Walk south towards the 1.8m high buttress that marks the margin of the Ardross Neck. Notice that the country rock is abruptly cut off at the margin of the neck, but that tuff veins pass out from it into the sandstone in which there are signs of local disruption for a few metres away from the margin. The tuff from the neck margin contains abundant blocks of sediment, including large masses of both sandstone and limestone, the latter believed to be the St Monans White Limestone. Further inwards it will be seen that the tuff is composed almost entirely of material of igneous origin and is cut by many calcite veins. Seawards the neck seems to end against the line of the Ardross Fault, but the contacts are not exposed.

Notice the NW–SE basalt dykes which cut the neck. These have been excavated by the sea near LWM but higher up the beach they stand up above the general level of the platform cut in the tuff (see Plate 4). The dykes contain large corroded anorthoclase xenocrysts, microphenocrysts of augite and pseudomorphs after olivine.

3. Ardross Neck: Ardross Fault and tuffisite

Westwards, a sandy bay interrupts exposures on the shore but, adjacent to the sand dunes and beyond the bay, the exposures still consist of tuff believed to be part of the same neck. Towards the south-western end of the Ardross Neck, tuff containing sedimentary material occurs on the south-eastern side of a trench marking the line of the Ardross Fault.

Unfortunately, much of the lower part of the beach here is

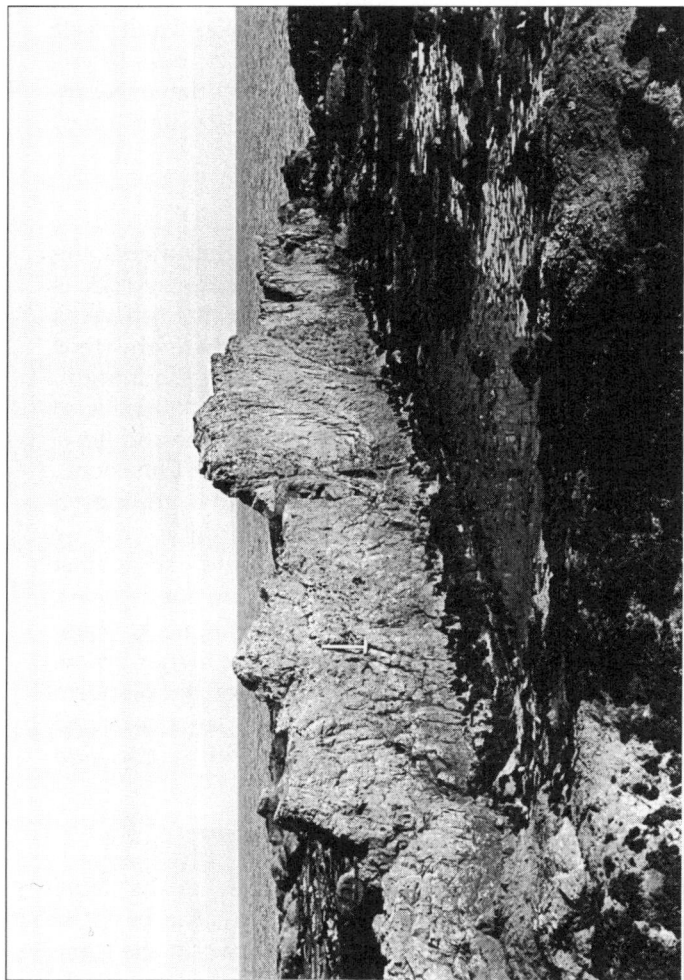

PLATE 4: Basalt dyke cutting the tuffs of the Ardross Neck, Elie. The tuff comprises almost exclusively igneous fragments; the dyke contains anorthoclase xenocrysts.

covered in loose boulders, but sufficient exposures occur between the boulders to indicate that there is a large area occupied by tuffisite and several small masses of basaltic breccia. The limits of the tuffisite cannot easily be determined. At LWM shales and sandstones outcrop in a minor anticline. These are believed to lie stratigraphically about 180m beneath the Ardross Limestones (Cumming 1936, p349).

4. Wadeslea Neck: eastern margin and tuff

Westwards from the Ardross Neck sediments outcrop for a short distance before the eastern side of the Wadeslea Neck, marked by another buttress, is reached. Note that the sediments are bent and buckled against the side of the neck and are cut by at least one small white trap dyke striking parallel to the neck margin. The visible outcrop of this neck is quite narrow at HWM and broadens seawards. It has a steep seaward face and a planed off top. The tuff in the greater part of the neck is of igneous origin, the 'dull green agglomerate' of Cumming (1936, p359), but there is a narrow marginal zone of tuffisite breccia.

5. Wadeslea Neck: sediments in the neck, white trap dykes

The western margin of the Wadeslea Neck is quite different from the eastern margin. It should be examined near LWM 140m east of Lady's Tower (499994). The precise boundary is hard to trace, there being a marginal tuffisite breccia zone in which quite large rafts of sandstone still retain approximately the dip and strike of the adjacent country rock. Note that the St Monans White Limestone crops out in one or two fairly prominent stacks near LWM and strikes into or indeed lies within the vent. To the north-east, away from the margin, the vent material is the more usual greenish tuff and rises as a 1.8m high buttress above the adjacent sediments and tuffisite breccia.

The sediments between the Wadeslea Neck and the Elie Ness Neck to the south-west are crossed by two NW–SE trending white trap dykes a few metres thick. These terminate to the north-west at the Elie Ness Neck while to the south-east one passes out to sea and the other ends part way down the shore. The country rock consists of sandstones and shale folded into a north–south anticline with dips of 20°–35°. To the south this fold is abruptly truncated by the Elie Ness Neck.

6. Wadeslea Neck: tuffisites

A series of small tuffisite necks cuts the eastern limb of the anticline at this locality and another lies on the anticlinal axis. These should be carefully examined. Most of them contain no igneous material; instead blocks of sandstone and shale up to 2.5m long are scattered at random through a mainly shaly matrix. Neck 2 contains white trap fragments in a tuffaceous matrix and Neck 8 has a tuffaceous matrix; it occupies a small fault and has weathered with a perfectly flat top so that it now looks like a small roadway on the beach. The majority of these small necks are probably related to the nearby Wadeslea Neck. They represent pipes up which little or no igneous material other than gas has passed and which are now filled with tuffisite composed of shaken-up and shattered country rock produced by high pressure gas action. The formation of such tuffisite was a common feature in the early stages of volcanic activity of the Carboniferous vents of Fife and the Lothians (Francis 1962).

7. Lady's Tower: Elie Ness Neck

The Elie Ness Neck is of considerable size, being at least 450m in diameter. The eastern margin of this neck is well exposed and may be examined in the small bay immediately east of Lady's Tower where the relationship between the neck and the country rock may be studied. At the southern end of the bay, there is considerable down drag of the beds as they approach the neck and the anticline referred to above is abruptly cut across, while at the northern end of the bay, the sediments are puckered and sharply bent and are veined with tuff. At HWM a white trap dyke packed with xenoliths and cut by many carbonate veins is well displayed. Drag of the sediments until they are almost vertical can also be seen.

8. Elie Ness Neck: tuff, bedded tuff and dykes

Near the eastern margin of the neck the tuff is poorly bedded and much faulted while light and dark coloured tuff may be seen to be in cross cutting relation to each other. Notice also the numerous, mainly small, short and irregular basalt dykes in this part of the neck. Much of the neck is occupied by bedded tuff dipping in the main concentrically at 15°–45° towards the centre of the vent. The bedded tuff is well displayed 90m south-west of Lady's Tower where the bedding is emphasised by distinct grain size and compositional variation. The coarser beds contain blocks of older tuff together with smaller masses of pyroxene and hornblende up to 10cm across (see Plate 5). Both large and small blocks of basalt are common and, together with the pyroxene and hornblende masses, are set in a fine tuff matrix which, as noted by Heddle (1901 vol. 2, p48), also contains irregularly distributed pyrope garnets locally known as 'Elie Rubies'. Cross bedding formed as part of base surge deposits is also present locally. Base surge deposits may be formed in Surtseyan type eruptions (Fisher and Schmincke 1984) thus agreeing with the opinions of Forsyth and Chisholm (1977) and Francis (1983) that the Carboniferous vents of East Fife are of this type. Some bands are fine-grained and darker in colour, owing to the presence of an appreciable amount of shale. Occasional sandy beds have also been recorded. Scattered limestone pebbles are conspicuous as they weather white or pale grey. Coal fragments, formerly wood and now cut by calcite veins, are also present though not common. A larger dyke, about 2m wide and striking north-south, is conspicuous 45m east of the lighthouse. In common with many of the other dykes, it contains corroded xenocrysts of hornblende, biotite and pink feldspar. A good example of faulting in the bedded ash can be seen a few metres west of this dyke.

PLATE 5: Bedded tuffs, Elie Ness Neck, Elie. The tuff comprises blocks of basalt and older tuff in which are scattered aggregates and crystals of pyroxene and amphibole and, very rarely, pyrope garnet or 'Elie Ruby'. Excursion 14, Location 8.

9. Elie Ness Neck: intrusion breccia plug, Ardross Fault

At the western end of the Elie Ness Neck and 50m north-west of the lighthouse examine a late plug of basaltic breccia measuring 25 × 18m. This passes steeply downwards through the bedded ash which is baked at the contacts. The plug consists of blocks of basalt up to 1.2m across set in a matrix of smaller basalt blocks, the interspaces being occupied by finer grained material and crystalline calcite. Bedded ash extends from this plug along the south-eastern side of Woodhaven to disappear at HWM. The abrupt disappearance of the Elie Ness Neck to the north-west may be due to the Ardross Fault. Lack of exposures prevents proof of this, but it seems quite probable (Cumming 1936, Francis and Hopgood 1970, p169).

10. Woodhaven: dolerite and limestones

A small mass of dolerite crops out in the centre of Woodhaven at HWM. It is cut by veins of tuffisite especially on the eastern side, while the western side is altered to white trap suggesting that the country rock may lie only a few metres away. Veins of carbonate up to 15cm wide traverse much of the rock. Just west of the dolerite, two limestones are exposed. That nearer the dolerite is recrystallised in part, probably as a result of metamorphism by the dolerite which has also affected the overlying shales to some extent. This limestone, which is crinoidal and at least 6m thick, is probably the St Monans White Limestone. The Woodhaven Limestone, about 1.5m thick, lies 40m higher in the succession. Both the limestone and the immediately overlying shales are fossiliferous yielding crinoids, 'Productus', 'Chonetes' and Composita. This limestone is believed to be equivalent to the Charlestown Main Limestone, Wilson (1966, p113) having discovered the Neilson Shell Bed fauna in the overlying shales. Almost all the sediments in Woodhaven therefore belong to the Lower Limestone Formation. Above this, a series of sandstones, shales and minor calcareous beds dips at 20°–50° to the north-west. The strike swings round to NW–SE on approaching the harbour wall. Note that these beds are cut by three branching white trap dykes which strike roughly E–W.

11. Woodhaven: tuffisite, dykes and margin of the Elie Harbour Neck

A basalt dyke appears near LWM in Woodhaven and a short distance further west cuts through a small tuffisite neck measuring 9 × 30m and aligned roughly along the length of the dyke. This neck appears to be composed entirely of brecciated sandstone and shale. By following this dyke west the well known Elie Harbour Neck is approached. Examine the eastern margin which is much faulted. Locally there is a limestone present which dips very steeply into the neck, as do the other beds. The tuff forms a buttress here too and stands above the level of the country rock.

12. Elie Harbour Neck: sandstone raft, bedded tuff

Excellent exposures of concentrically dipping, well bedded tuff may be examined within the Elie Harbour Neck, the centre towards which they dip lying beneath the position of a large sandstone raft, 75m south-east of the warehouse at the harbour. The tuff, in which dips are as high as 75°, is similar to that in the Elie Ness Neck. It contains large basalt blocks, some of which are vesicular, and also small pieces of white-weathering limestone, some of which are crinoidal. Perhaps the most interesting feature within the neck is the relationship of the large sandstone raft to the adjacent tuff. The sandstone itself is not unusual, though it is locally pink due to relatively high concentrations of almandine garnet, and it contains minor shale bands and a thin coaly bed. When the contact between this sedimentary raft and the tuff is examined, however, it is apparent that there is a passage between the one rock type and the other. The intermediate rock has the appearance of a coarse sandstone or grit, but contains tiny chips of pale, greenish-grey, altered basalt or white trap. It seems likely that gas action has broken down the sandstone at the margin of the raft and introduced small fragments of tuff or lava, so that a sharp boundary between the two rock types no longer exists. An unaltered basalt dyke appears within the bedded ash and passes close to the sandstone raft before dying out a short distance west. The bedded tuff can be followed to the western limit of the exposures without finding any sign of the edge of the neck being near. A few exposures of tuff occur within the harbour at the foot of the quay and indicate, as Cumming (1928, p135) pointed out, that the neck was much larger than was originally thought by Geikie (1902, pp244–5).

Rejoin the bus at the harbour and return to St Andrews by the same route as taken on the outward journey.

References

CUMMING, G. A., 1928. The Lower Limestones and associated volcanic rocks of a section of the Fife coast. *Trans. Edinb. Geol. Soc.* **12**, 124–40

————, 1936. The structural and volcanic geology of the Elie–St Monance district. *Trans. Edinb. Geol. Soc.* **13**, 340–65.

FISHER, R. V. and SCHMINKE, H. U., 1984. *Pyroclastic rocks.* Springer.

FORSYTH, I. H. and CHISHOLM, J. I., 1977. The geology of East Fife. *Mem. Geol. Surv. Gt. Br.*

FRANCIS, E. H., 1962. Volcanic neck emplacement and subsidence structures at Dunbar, southeast Scotland. *Trans. Roy. Soc. Edinb.*, **65**, 41–58.

————, 1983. Magma and sediment II. Problems in interpreting palaeo-volcanics buried in the stratigraphic column. *Jour. Geol. Soc. Lond.*, **140**, 165–83.

———— and HOPGOOD, A. M., 1970. Volcanism and the Ardross Fault, Fife. *Scot. Jour. Geol.*, **6**, 162–85.

GEIKIE, A., 1902. The geology of eastern Fife. *Mem. Geol. Surv. Scot.*

HEDDLE, M. F., 1901. *The mineralogy of Scotland.* Vol. 2. Douglas, Edinburgh.

WILSON, R. B., 1966. A study of the Neilson Shell Bed, a Scottish Lower Carboniferous marine shale. *Bull. Geol. Surv. Gt. Br.* No 24, 105–30.

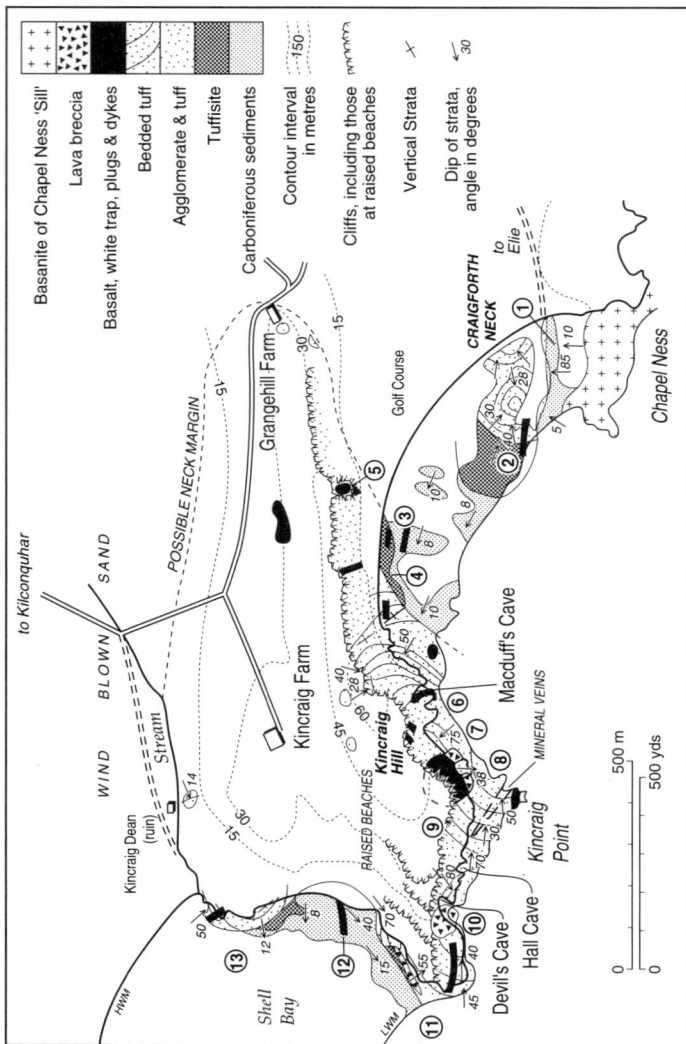

MAP 20: Kincraig - Chapel Ness, Elie.

Legend:
- Basanite of Chapel Ness 'Sill'
- Lava breccia
- Basalt, white trap, plugs & dykes
- Bedded tuff
- Agglomerate & tuff
- Tuffisite
- Carboniferous sediments
- Contour interval in metres
- Cliffs, including those at raised beaches
- Vertical Strata
- Dip of strata, angle in degrees

to Kilconquhar

WIND   BLOWN   SAND

POSSIBLE NECK MARGIN

Kincraig Dean (ruin)

Kincraig Stream

Grangehill Farm

Kincraig Farm

Golf Course

CRAIGFORTH NECK

to Elie

Chapel Ness

Macduff's Cave

Kincraig Hill

MINERAL VEINS

Kincraig Point

RAISED BEACHES

Hall Cave

Devil's Cave

Shell Bay

HWM

LWM

500 m

500 yds

240

Excursion 15

Kincraig (half day)

OS 1:50,000, sheet 59
GS One inch, sheet 41
Excursion map 20.

WALKING DISTANCE: 3km of track and sandy beach; 0.8km of arduous rock scrambling – see warning below.

PURPOSE: The main purpose of the excursion is to examine an exceptionally well exposed Carboniferous volcanic neck, Kincraig Neck, which is exposed amid fine coastal cliff scenery west of Elie on the Firth of Forth. Within it the relationships of plugs, dykes and tuff to each other and to the country rock can clearly be seen in the cliffs and in the wavecut platform.

WARNINGS:

1. Access to the shore between Localities 6 and 10 is restricted by the tide to 2.5 hours on either side of low tide.

2. Since the route round the foot of the cliff involves several climbs and descents by means of chains and footholds cut in the rock, those less certain of their footing are recommended to follow the good cliff-top path from Locality 5 to beyond Locality 10 where they may rejoin the shore route.

ROUTE: Follow A915 south from St Andrews to Largoward, then B941 through Kilconquhar to join A917 1.5km north of Elie; follow A917 into the town as far as the parish church and turn right at the T-junction. Now travel westwards through Earlsferry for 1km to a car park beside the golf course, where the bus should remain. From here the excursion should be followed to Locality 10 before returning along the cliff-top path which gives an excellent 'aerial' view of the exposures. Alternatively, if more time is available, follow the excursion to Locality 13 before either returning or arranging to be picked up at the Shell Bay Caravan Site.

1. *Chapel Ness Basanite Intrusion: margin of Craigforth Neck; bedded tuff and dyke*

Follow the track westwards across the golf course for 0.5km to the shore at the margin of the Craigforth Neck lying south-east of the main Kincraig Neck. One hundred and forty metres south-west lies the Chapel Ness Basanite Intrusion. It forms a step-like feature across the beach and displays well developed columnar jointing. The rock, which is very fine grained, is grey when fresh, but takes on a faint purple hue on weathering. It is amygdaloidal towards the base and also contains a 1.5m thick dyke displaying good horizontal columns. The underlying sediments are gently dipping and are separated from the intrusion by an extensive breccia zone containing large rafts of country rock in a matrix of disrupted sediment.

A careful examination of these rocks will show that white trap veins from the base of the intrusion extend into the breccia zone and the country rocks. Tuff and agglomerate crop out low on the shore and are separated from the intrusion by the breccia zone. These features are more reminiscent of a neck intrusion. On its eastern side the intrusion is sill like, being generally in concordance with the sediments of the Limestone Coal Formation. Forsyth and Chisholm (1977, p188) suggest that it is indeed a neck intrusion, but one which extends from the neck into the sediments as a sill on the eastern side.

One hundred and forty metres to the north of the intrusion the gently dipping country rock is bent down steeply (up to 85°) into the margin of the Craigforth Neck. Here the country rock is hardened, thus standing up above the surrounding rocks, and is cut by many calcite veins. Notice that the neck, which here contains unbedded, very shaly tuff, is also cut by many calcite veins for the first few metres inwards from the margin. On passing north-westwards across the shore, well bedded tuff is reached and in this it is possible to make out concentric dips associated with at least three separate centres within 140m of the neck margin. A 6m thick basalt dyke is well exposed low on the shore.

2. *Craigforth Neck: tuff and sediments, tuffisite*

North-west from the bedded tuff centres there is an outcrop of little disturbed sandy beds striking approximately E–W, apparently lying within the neck and dipping parallel to the tuff. The remainder of the neck is occupied by tuffisite composed mainly of sediment, apparently broken up in place by gas action with beds of sediment (up to 20cm thick) still in place. The western margin of this tuffisite is steeply inward dipping and similar to the margin on the south side of the vent.

3. *Basalt dyke with white trap margins*

For the next 275m along the shore to the WNW, sandstone ridges dipping west at 8°–10° protrude from the sand. At one of these ridges, lying 27m from the margin of the Kincraig Neck, examine carefully an E–W basalt dyke, now altered to white trap at its margins, especially where it divides to form thin stringers running into the country rock. Where the dyke is about 3m thick, the centre is little altered. Calcite veins in the white trap run parallel to the length of the dyke.

4. *Margin of Kincraig Neck: intrusion breccia; bedded tuff*

At this locality note that the south-eastern side of the neck has a sheath of marginal intrusion breccia or tuffisite which, although at least 25m wide at the top of the beach, tapers to only a few metres at LWM. There is little sign of forcible displacement of the country rock, rather the beds are simply truncated by the intrusion breccia with slight bending into the neck for a few metres only. Examine the intrusion breccia

FIGURE 14: Kincraig Neck, Elie, seen from the golf course. Notice the well bedded tuff dipping towards Centre 1 within the neck. The dark distant cliff is columnar-jointed basalt, also within the neck.

which consists mainly of fairly small fragments of shaly material cut by sporadic, highly irregular white trap masses. It contains blocks up to several metres across of both sandstone and shale. The large blocks are well exposed at the top of the beach north of the white trap dyke at Location 3. Isolated fragments of white trap, only a few centimetres across, also occur in the shaly matrix. Intrusion breccia of this type is generally believed to be emplaced at an early stage in the formation of the Fife necks, but this is apparently not always the case; here the breccia invades not only the country rock, but also the bedded ash within the neck.

On passing north across the intrusion breccia, dull grey-green bedded tuff is encountered and the bedding within this is splendidly displayed in the cliffs of Kincraig Hill 75m further west. On the shore, steeply dipping tuff is broken by faults every few tens of metres with changes of strike across the faults. Notice that the tuff varies in grain size from fine volcanic dust to lava fragments centimetres across with larger lava fragments scattered throughout. Pieces of older bedded tuff and sediment are not common. This tuff is invaded by small tuffisite dykes and also by small basalt masses, one far down the shore and only accessible at low tide, another forming the crag at Location 5.

The bedding in the tuff becomes indistinct in places (and may disappear either gradually or abruptly, as at faults), but where it can be traced for any distance, it dips towards a series of centres though nowhere is a complete circle round such a centre seen. Four centres can be recognised in the Kincraig Neck, though the tuffs associated with them occupy less than half the known area of the neck. The first of these lies approximately 75m west of MacDuff's Cave and the tuffs of Kincraig Hill dip towards it.

5. West end of golf course: basalt plug

This locality may be examined on the return to the bus along the cliff top path.

A small basalt plug 15m across forms a prominent crag 255m north-east of Location 4 and on examination will be seen to consist of two small masses showing good columnar jointing. Stringers of basalt run out from it into the adjacent tuff, but there is little sign of alteration in the tuff. Veins of calcite are widespread within the plug.

6. Macduff's Cave: bedded tuff, fault and dyke

At the gully in the cliff called MacDuff's Cave, where the first chains occur on the shore path, note that the bedded tuff is cut by a fault along which an impersistent, up to 20cm thick, basalt dyke has been emplaced. The dyke is found on the west side of the gully. Beyond the fault the dip is variable, but in general concentric about the same focus. After passing the prominent headland 75m south-west of the cave, notice that the bedding becomes indistinct, though it can still be discerned high in the cliff.

7. Main plug of Kincraig Neck: columnar jointing

High on the cliff in the northern corner of the next bay, note three small irregular dykes cutting the tuff. The rocks on the shore here should be examined carefully. The bedded tuff dips at 75° to the NNW and contains irregular blocks of similar, older, bedded tuff up to 2m across. This blocky tuff continues west, but becomes brecciated for the last few metres before the junction with the main plug is reached. The plug is in two parts, probably continuous, but divided by a grass covered gully, columnar jointing being well displayed in both parts. In the eastern part, individual columns are as much as 20m long, sweeping unbroken up the cliff face. The inclination of the columns suggests that the lava which cooled to form them occupied a conical depression (Forsyth and Chisholm 1977, p186), presumably overlying the feeder pipe (cf. Whyte 1966, Fig.3, p111). Forsyth and Chisholm suggest, however, that it overlies a collapse breccia.

8. Kincraig Point: southern margin of basalt plug, intrusions and lava breccia

The next chain at the western side of the bay allows one to pass a small headland of tuff (which may be skirted at low tide) abutting against the plug. Beyond it is a wide wave cut platform in bedded tuff, which although similar to that seen further east, dips to the south and south-east at 30°–50° and is therefore associated with a second centre. When this tuff is traced towards the western part of the plug, a new rock type is met: a breccia of lava blocks up to 60cm across which forms an intermittent screen between the plug and the bedded tuff. A further small part of this screen occurs against the eastern part of the plug, but is largely obscured by fallen columns and other debris. At the top of the next chain, where the basalt is in contact with the bedded tuff, the columnar jointing gives way to a platy jointing some metres from the tuff (see Plate 6). When the contact between the lava breccia and the plug is examined at the base of the chain, it will be seen that tuff has been injected along the contact and thus provides evidence

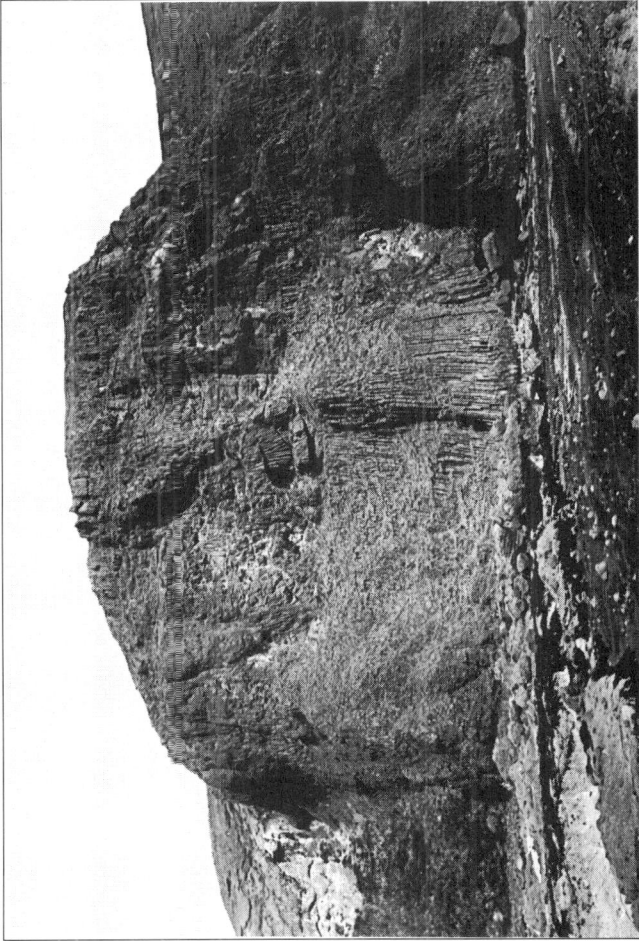

PLATE 6: Columnar-jointed basalt in the main plug, Kincraig Neck, Elie. The wave-cut platform in the foreground truncates bedded tuff dipping towards Centre 2 within the neck. Excursion 15, Location 8. (Geological Survey Photo, Crown Copyright Reserved).

of a later phase of igneous activity. This tuff dyke, about 0.3m wide, is purely local.

Notice that the western plug has a break in its columnar jointing about halfway up (this is more clearly seen with the aid of binoculars) and it appears that at least two intrusions are represented in the visible part of the basalt mass (Geikie 1902, p12).

9. *Western margin of main plug: mineral veins; Hall Cave to Devil's Cave*

That the lava breccia is later than the plug can be demonstrated by following their junction westwards round the edge of the plug and then for 15m up a gully running part way up the cliff from the shore. In this gully the breccia can be seen to abut against the columns and several of these have been wedged loose so that they now lie in the breccia. Seams of tuff also penetrate between the columns which are essentially in their original positions. In the cliff beneath an old gun emplacement, about 25m west of the plug margin, a mass of columns of lava measuring 1.5 × 0.9m lies enclosed in the agglomerate, a clear indication of their age relationship. A late stage discharge of gas at the western margin of the plug appears to have broken up the columnar basalt, the gas penetrating between the columns and prizing some of these away from the main mass. To the west and south the same gas action has disrupted the bedded tuff and, for some metres west of the main plug, the outcrops consist of unbedded tuff or agglomerate in which lie broken up columns of basalt.

On the wave cut platform the sea has, in several places, cut back along joints in the bedded tuff. A small E–W dyke is exposed on the shore platform at Kincraig Point, but dykes are generally rare on this part of the shore. Minor mineralisation is present in veins in the bedded tuff, small amounts of pyrite being present in a gangue composed of barytes, calcite and quartz (amethyst). The bedded tuff of Kincraig contains nodules composed of mineral aggregates or single crystals; hornblendite, pyroxenite, biotite and olivine have all been recorded. The bedding of the tuff continues to dip towards Centre 2 for another 180m west from the main plug. At the western end of the wave cut platform, it is necessary to use chains once more to reach Hall Cave and to continue west from it.

Just east of Hall Cave the bedded tuff of Centre 2 is truncated at a gully by that of Centre 3 and thereafter the dips in the bedded tuff are about 80° south-west towards a centre lying about 90m south-west of Devil's Cave. Notice that here too the tuff alternates between coarse- and fine-grained beds and contains large isolated blocks.

10. West side of Devil's Cave: basalt dykes, erratic of augen-gneiss, bedding in tuff

On the western side of Devil's Cave a mass of lava breccia 60m across may be examined. It is apparently plug-like, but has irregular non-vertical margins. A small number of very irregular basalt dykes cuts the eastern part of this mass and two more occur in the bedded tuff a few metres south of the breccia; none is more than 1m across. Westwards, the lava breccia passes by gradation into unbedded tuff. In a small bay 35m south-west of Devil's Cave lies a large block of pink augen-gneiss measuring 1.2 × 1.0 × 1.0m and presumably an erratic block deposited by the Forth Glacier during the Devensian glaciation. West from here to the point at Shell Bay, the bedded tuff dips ESE at 40°–50°. It is cut by a 1m thick dyke and associated calcite veins. At the headland 75m south-west of the dyke, the bedded tuff contains examples of low angle cross bedding and graded bedding. These features may indicate a base surge deposit similar to that found at Elie Ness (Excursion 14, Location 8).

Return to the bus along the cliff top or alternatively carry on to Locations 11, 12 and 13.

11. Western margin of Kincraig Neck: tuffisite and lava breccia

Passing north-east and then north from the point at Shell Bay, the margin of Kincraig Neck is reached. Parallel to it is an area of coarse lava breccia, 10 × 135m, presumably the product of a late centre of activity comparable to that at Devil's Cave. At the neck margin, which strikes NE–SW parallel to the coastline, is a hardened sandstone ridge which is succeeded inwards by tuffisite and then, locally, the lava breccia followed by the bedded tuff, now dipping to the south-east at 55°–75°. Just before the neck margin reaches HWM, the dip of the bedded tuff has changed to 70° to the south-west, thus continuing the swing in strike round Centre 3.

12. *Shell Bay: country rock and dykes*

Examine here sandstones and shales of the Limestone Coal Formation, lying on the western side of the neck, which have a general westerly inclination of 10°–12° modified by gentle flexuring. These sediments are cut by several tuffisite dykes and a white trap dyke.

13. *Shell Bay; neck margin*

The margin of the Kincraig Neck reappears towards the

FIGURE 15: The raised beaches on Kincraig Hill, Elie seen from Shell Bay. These lie at about 4, 11.5, 22 and 24.5m OD and mark stages of stillstand during isostatic uplift of the area following deglaciation.

northern end of Shell Bay. Striking NW–SE at first it swings round to N–S before disappearing under the sands of the bay. Along most of its length hereabouts, there is a sheath of tuffisite which in one place tongues south into the country rock for 25m. As in previous occurrences the tuffisite consists of shaly and sandy blocks set in a shaly matrix with spots and streaks of white trap. The bedded tuff within the neck strikes parallel to the margin and dips at high angles throughout. When last seen at the mouth of the burn, the dip is to the south-east, i.e. towards Centre 4. In the same exposures the bedded tuff is cut by small, impersistent and twisting basalt dykes. In an isolated exposure south of Kincraig Dean the tuff dips almost due west at 14°, implying a further swing round the fourth centre.

Raised beaches

At the western end of Kincraig Hill are four very well preserved raised beaches of Late-glacial and Postglacial age. These can be clearly seen from Shell Bay. They are cut into the soft, easily eroded, bedded tuff at approximately 3.6m, 11.5m, 22m and 24.5m above sea level (Cullingford and Smith 1966, p40) and are analogous to the beach beneath them on the present day coastline, e.g. at Location 9. The origin of such beaches is discussed on pp 69-71.

Elsewhere on Kincraig Hill there are few exposures, but in the remains of the army camp on top of the hill, bedded tuff, now deeply weathered, occurs in trenches and other excavations. Basalt with well developed columnar jointing outcrops in a rocky crag 450m east of Kincraig Farm, presumably forming part of another plug. Scattered exposures of tuff continue eastwards as far as Grangehill Farm, but the location of the neck margin is unknown.

Return along the cliff top path to the bus, or rejoin the bus at the caravan site.

References

CULLINGFORD, R. A. and SMITH, D. E., 1966. Late-glacial shore-lines in Eastern Fife. *Trans. Inst. Brit. Geogr.*, **39**, 31–51.

FORSYTH, I. H. and CHISHOLM, J. I., 1977. The geology of East Fife. *Mem. Geol. Surv. Gt. Br.*

GEIKIE, A., 1902. The geology of Eastern Fife. *Mem. Geol. Surv. Scot.*

WHYTE, F., 1966. Dumbarton Rock. *Scot. Jour. Geol.*, **2**, 107–21.

East Lomond
(half day)

OS 1:50,000 Sheets 58, 59
GS One-inch Sheet 40
Excursion map 21.

WALKING DISTANCE: 5km of track and grassy hill walking.

PURPOSE: To examine (1) part of the great Midland Valley Quartz-dolerite Sill exposed over much of the East Lomond area, including contact phenomena at its base and top; (2) Lower Carboniferous sediments adjacent to the Sill; (3) the East Lomond volcanic neck with its agglomerate and olivine-dolerite plug, the latter forming the highest part of the hill; and (4) the view from the summit of the geology of much of Fife and on a clear day across to the Lothians to the south and to the Sidlaws and the Scottish Highlands to the north and west.

ROUTE: Proceed by A91 to Cupar then south-west on A92 along the foot of the scarp formed by the Midland Valley Sill through Pitlessie and Kettlebridge to the roundabout at the junction with A914 and A912. Now follow A912 north-westwards for 1.5km to the signposted road on the south-west leading to the picnic site on the East Lomond. This is a single-track but good road with passing places and there is ample room to turn the bus at the picnic site (2km) and TV tower (252058).

Send the bus back to A912 and then through Falkland and up the unclassified road from Falkland to Leslie as far as the

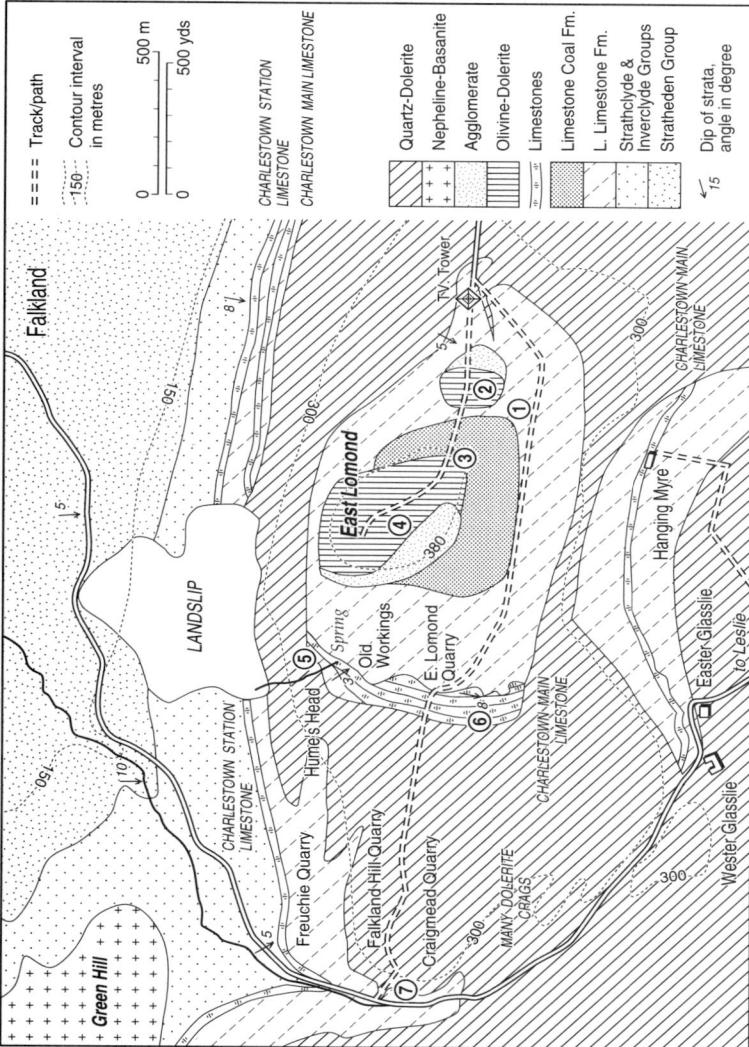

MAP 21: East Lomond.

FIGURE 16 The East Lomond seen from above Hanging Myre Farm. The summit is composed of a very late Carboniferous to early Permian olivine-dolerite volcanic plug, while the shoulder on the left comprises agglomerate.

Craigmead car park (227063). If a bus is not being used it is best to drive to the Craigmead car park direct and walk eastwards for 2km, using the hill road, to the TV tower at the picnic site.

1. Hanging Myre Farm: view of the East Lomond

Six hundred and fifty metres west of the TV tower the hill road passes north of Hanging Myre Farm (246053), a site of some historical interest, with Lower Limestone Formation sediments indifferently exposed in the scarp behind the farm house. The sediments are capped by the quartz-dolerite sill, here no more than 30m thick in contrast to its 60m thickness above Falkland. An old kiln in the vicinity of the farm indicates that the Charlestown Main Limestone was at one time worked there. The farm stands on the site of an unsuccessful lead and silver mine, opened about 1783 and said to have worked galena in a NE–SW vein. Of it Heddle (1901, pp18, 19)

255

stated, 'The content of silver and other wonders are the statements for the most part of interested parties . . . They do not now even yield lead.'

From the hill road the appearance of the East Lomond is of a rounded summit with a marked shoulder on the western side. The summit is formed of olivine-dolerite and the shoulder of agglomerate.

2. *Small neck*

The path to the summit of the East Lomond runs west from the TV tower and first crosses a very poorly exposed small neck containing an olivine-dolerite plug and agglomerate, outcrops only being seen in ruts in the path and among the heather on either side.

3. *East Lomond: agglomerate and olivine-dolerite*

On the path, about 650m from the TV tower, the margin of the main East Lomond neck is reached, but exposures here are poor. This neck and the basanite neck of the West Lomond are thought to cross-cut the Midland Valley Sill but the relationship cannot be proved on the ground. Cameron and Stephenson (1985, pp110,119) discussed the likely sequence of igneous activity in the Midland Valley and noted the respective age dates for such quartz-dolerite sills at about 295–290My and of such basanites at 290–280My, i.e. the sills as late Westphalian to early Stephanian and the necks as a little younger, late Stephanian to early Permian.

Continue up the path to 150m beyond a stile in the wall, which crosses the path, until a 1m high trigonometric station is seen 2m south of the path. By walking a short distance south-west from the trig. station small outcrops of agglomerate can be seen, and display the soft and friable nature of the rock. The agglomerate is pale olive-grey in colour and contains fragments of very rotten lava up to 1.5cm in diameter. One reasonably fresh fragment examined by Walker and Irving (1928, p6) proved to be of olivine-dolerite.

Now examine the prominent crags between the trig. station

and the top of the East Lomond. These are of olivine-dolerite showing irregular columnar jointing, described by Irving (1929) as radiating from the centre of the mass. Many exposures of this rock occur all round the top of the hill and some have yielded quartz xenocrysts (Walker and Irving 1928, p5). In hand specimen the rock is dark green to black in colour and olivine can usually be seen. Vesicles with green chlorite also occur and the rock is generally very fresh, a feature in marked contrast to the normally deeply weathered quartz-dolerite of the sill, e.g. in Craigmead Quarry, Location 7. The spheroidal weathering so common in the quartz-dolerite is also absent.

4. Summit of the East Lomond: regional geology

Many of the hills and peaks seen from the summit can be identified from the direction indicator. To the west and projecting 120m above the very obvious feature formed by the Midland Valley Sill lies the West Lomond. This is a neck occupied mainly by a basanite plug and surrounded by sediments. To the south the ground falls away towards the syn-clinal Fife Coalfield, centring on Leven and including rocks of Upper Carboniferous (Westphalian) age belonging to the Coal Measures. This structure is continued on the south side of the Firth of Forth in the Lothian Coalfield, immediately to the east of the Pentland Hills. These hills comprise Lower Old Red Sandstone volcanic rocks brought into faulted contact with the coalfield by the large Pentland Fault. On a clear day the Carboniferous volcanic necks of Arthur's Seat and the Castle Rock in Edinburgh are visible as is the teschenite sill of the Salisbury Crags.

In Fife, to the east notice the volcanic necks of Largo Law and Kellie Law. To the north-east the ground falls away to the Howe of Fife, underlain by almost horizontal Stratheden Group sediments which, along the south side of the Howe, pass up into Inverclyde Group and very thin Strathclyde Group rocks where they are protected by the Midland Valley Sill, which extends from the Lomond Hills almost to Cupar.

257

North of the Howe and Stratheden the lavas of the Ochil Volcanic Formation of the Lower Old Red Sandstone form the North Fife Hills, a continuation of the Ochil Hills, and dip at about 20° SE under the horizontal Stratheden Group, approximately equivalent to the Upper Old Red Sandstone. Beyond these hills to the north lie the Sidlaw Hills, also composed of Ochil Volcanic Formation lavas, but this time dipping north-west on the other side of the Sidlaw Anticline. In the far distance many peaks of Highland Dalradian rocks stand up beyond the Highland Boundary Fault.

During the late Devensian glaciation ice movements in the area were from west to east with the main Forth ice lying to the south of the Lomond Hills and east-flowing ice also moving down Stratheden. The Howe of Fife is covered by an extensive spread of fluvio-glacial sands and gravels which are exploited in the Ladybank-Collessie area. To the north, too, ice moved eastwards in the Carse of Gowrie between the Sidlaws and the North Fife Hills.

The summit of the East Lomond is the site of a hill fort with two encircling ramparts, best seen to the north-west of the summit, and of 'late first millenium BC – early first millenium AD' age (Walker and Ritchie 1987, p165). The direction indicator on the summit stands on a bronze-age cairn.

5. Hume's Head spring: *top of quartz-dolerite sill; baked sediments*

Continue west and descend from the summit down the fairly steep slope of olivine-dolerite crags to reach the steep grassy slope on agglomerate. An E–W wall passing 180m north of the summit extends for another 1.5km to the west. Keep about 65m south of this wall and follow it for 275m from the foot of the steep agglomerate slope to the spring at Hume's Head where fossiliferous baked shales crop out. By following the stream that flows north out of the spring for 20m, the top of the Midland Valley Quartz-dolerite Sill is reached. The shales must therefore lie only a few metres above the sill. A long line of old workings in the Charlestown Main Limestone can be followed SSW from the spring for 550m to the East Lomond

Quarry beside the old road leading westwards to Falklandhill Quarry. The limestone is exposed 225m south of the spring and is recrystallised, buff weathering, partly fine grained and partly dolomitic. One hundred and ten metres further on recrystallised limestone is again exposed; the fossils in it have weathered out leaving moulds. The remaining old workings show only loose sandstone blocks and small exposures of baked shale.

6. East Lomond Quarry: Charlestown Main Limestone and kiln

Old workings, some of them flooded, extend for some 200m south of the road. One hundred and sixty metres to the south about 1.2m of muddy limestone is exposed and contains corals, but is recrystallised. There is a Fife Ranger Service Industrial Heritage Trail in these old workings including a restored kiln at the southern end of the workings. If time permits the trail is well worth following giving an indication as it does of the economy involved. The shortage of limestone over much of Fife led to workings here at over 350m above sea level in the Charlestown Main Limestone by a team believed to be of six men, two boring and blasting, one carting, two breaking and loading the kiln and one drawing the lime and supplying the customers.

7. Craigmead Quarry: base of the quartz-dolerite sill

Now walk westwards along the moorland road for just over 1km to Craigmead Quarry. The road is underlain by quartz-dolerite of the Midland Valley Sill which both forms crags and from time to time crops out in the bed of the road. Part way down the steep hill and just before reaching the tarred road turn south into Craigmead Quarry. There the following succession at the base of the sill is exposed:

Columnar-jointed, spheroidal-weathering quartz-dolerite with fine-grained base	4.5m
Baked sandstone wedging into the sill	0.6
Dolerite from the sill	0.6
Sandstone, slightly baked at the top	1.0 +

The sill is chilled to a fine-grained, dense, black, basaltic rock for a few millimetres from the contact. The contact is transgressive and the horizontally bedded sandstone is baked to a quartzite for a short distance below the sill. Walker (1958, p113) found evidence in this quarry for very slight mobilisation or rheomorphism of the Carboniferous sediments at the base of the sill. The mobilised rock, which is believed to have originated in shalier laminae within the sandstone, lies between the sill and the sandstone and is 1mm to 3mm thick. Under the microscope Walker described it as invading the sandstone transgressively, but not penetrating the edge of the sill. This phenomenon is not visible in the field on account of its small scale.

Falklandhill Quarry, on the northern side of the hill road, displays the same major features but is more deeply weathered.

Two hundred metres north along the Falkland road is a now overgrown sandstone quarry, at one time worked for building stone. Opposite a gate in the roadside wall and 50m further downhill is Freuchie Quarry. In it shales and sandstone underlying the Charlestown Station Limestone are exposed. These too lie beneath the Midland Valley Sill. The limestone itself is poorly exposed and like the shales is not very fossiliferous. Forsyth and Chisholm (1977, p60) correlated the limestone with the Hurlet Limestone which marks the base of the Lower Limestone Formation.

Rejoin the bus at Craigmead car park and return to St Andrews by retracing the outward route.

References

CAMERON, I. B. and STEPHENSON, D., 1985. The Midland Valley of Scotland. *Br. Reg. Geol. (3rd Ed.)* HMSO.

FORSYTH, I. H. and CHISHOLM, J. I., 1977. The Geology of East Fife. *Mem. Geol. Surv. Gt. Br.*

HEDDLE, M. F., 1901. *The mineralogy of Scotland*. David Douglas, Edinburgh.

IRVING, J., 1929. The Carboniferous igneous intrusions of north-eastern Fifeshire. *Unpublished Ph.D. thesis, University of St Andrews.*

WALKER, B. and RITCHIE, J., 1987. *Exploring Scotland's Heritage, Fife and Tayside.* Royal Commission on the ancient and historical monuments of Scotland. HMSO.

WALKER, F., 1958. Dolerite-sandstone contact phenomena on the East Lomond, Fife. *Trans. Edinb. Geol. Soc.* **17**, 113–16.

————— and IRVING, J., 1928. The igneous intrusions between St Andrews and Loch Leven. *Trans. Roy. Soc. Edinb.* **56**, 1–17.

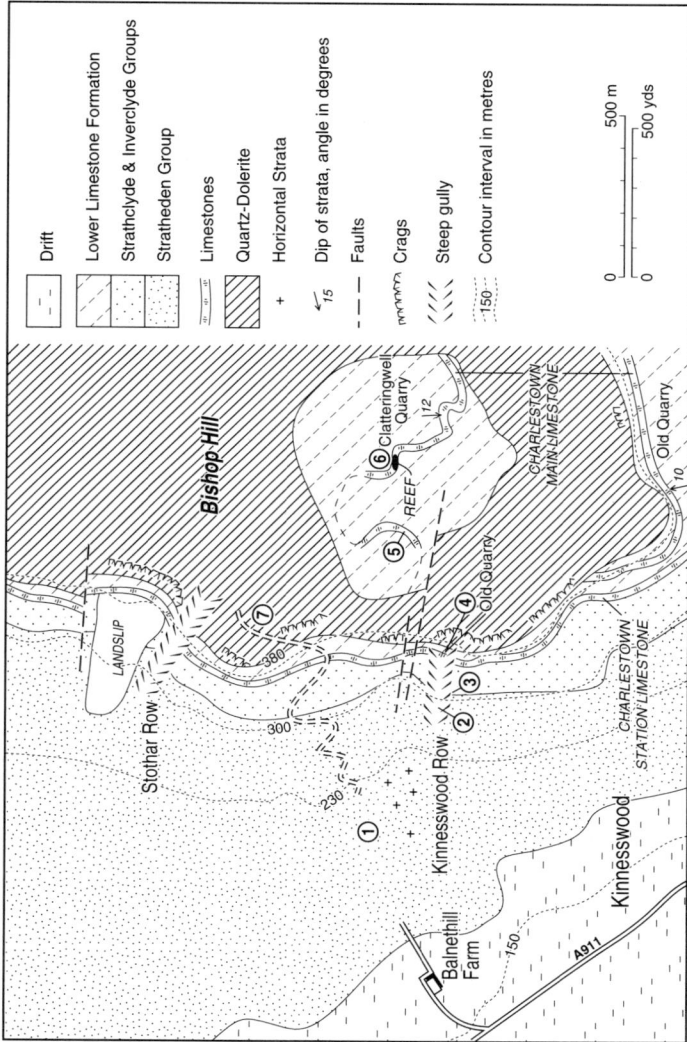

MAP 22: Bishop Hill.

Legend:

- Drift
- Lower Limestone Formation
- Strathclyde & Inverclyde Groups
- Stratheden Group
- Limestones
- Quartz-Dolerite
- + Horizontal Strata
- ↙15 Dip of strata, angle in degrees
- Faults
- Crags
- Steep gully
- 150 Contour interval in metres

0 ───── 500 m
0 ───── 500 yds

Bishop Hill

LANDSLIP

Stothar Row

Kinnesswood Row

Balnethill Farm

Kinnesswood

A911

Clatteringwell Quarry

REEF

CHARLESTOWN MAIN LIMESTONE

Old Quarry

CHARLESTOWN STATION LIMESTONE

Old Quarry

Bishop Hill
(half day)

OS 1:50,000 Sheet 58
GS One-inch Sheet 40
Excursion Map 22.

WALKING DISTANCE: 3.5km of grassy hill walking with steep ascent and descent of 245m in 800m.

PURPOSE: To examine (1) beds of the Knox Pulpit Formation of the Stratheden Group; (2) the Kinnesswood Formation of the Inverclyde Group; and (3) the sediments of the undifferentiated Strathclyde Group of the Carboniferous (Paterson and Hall, 1986). The last of these is less than 30m thick on Bishop Hill, in striking contrast to a thickness of over 2000m in the Anstruther area 38km to the east. The highest strata exposed in the area belong to (4) the Lower Limestone Formation of the Carboniferous and these are intruded by (5) the quartz-dolerite of the Midland Valley Sill (also seen on Excursion 16). (6) The summit of the scarp on Bishop Hill affords an excellent view of the regional geology including the eastern Ochil Hills and the Loch Leven area.

ROUTE: By A91 from St Andrews through Cupar and west across the Howe of Fife and Stratheden as far as the crossroads 3.5km north-east of Milnathort, then follow B919 south-eastwards for 2.5km to join A911 leading after 800m to Easter Balgeddie. The road to Balnethill Farm lies at the south-eastern end of the village and the bus should be left at the main road.

FIGURE 17: Bishop Hill, Loch Leven from Balnethill farm. The summit is capped by quartz-dolerite of the Midland Valley Sill. The steep face is composed of Carboniferous sediments passing down into sediments of the Stratheden Group. Kinnesswood Row is the gully on the right.

Walk up the farm road (ask permission at the farm) through the farmyard and on for 100m to where there is an excellent view of the scarp face of Bishop Hill. The top of the scarp is composed of columnar-jointed quartz-dolerite of the Midland Valley Sill and beneath this, above the prominent gully of Kinnesswood Row, can be seen the outcrop of pale sandstones overlying the Charlestown Station Limestone. Much of the lower slope is scree covered and the farmhouse itself stands on glacial till.

1. Balnethill Farm: Knox Pulpit Formation, Stratheden Group

One hundred metres up the track behind the farm buildings enter the field on the right before walking east to a low grassy mound in the same field near the base of which soft, pale buff, fine- to very fine-grained cross-bedded sandstone can be seen. This belongs to the Knox Pulpit Formation (Chisholm and Dean 1974). Continue up the field to the gate, then walk south across the steep hillside, over more cross-bedded sandstone, to a small gate in the wall, beyond which a 4.5m high waterfall can be seen. Here too the rock is cross-bedded, cream coloured, friable, fine-grained sandstone of the Knox Pulpit Formation. Occasional medium-grained sand grains occur and include well rounded millet-seed grains. Hall and Chisholm (1987, p203) identify these, on the basis of their sedimentary structures, as cross-bedded aeolian dune sands in which the paleowind was from the east. Fifty-five metres south of the waterfall is Kinnesswood Row in which mudflows occasionally occur owing to the presence of soft silty and muddy beds within the Kinnesswood Formation higher up the gully (Chisholm and Dean 1974).

2. *Kinnesswood Row: Kinnesswood Formation, Inverclyde Group*

In Kinnesswood Row the dip of the strata is about 5° E. As a result of weathering the sandstone beds form ledges on the hillside while the softer siltier beds have been eroded back between them. The sandstones, which are often cross-bedded and feldspathic with currents predominantly from the northwest, are limonitic brown in colour and contain sporadic gritty bands and mud pellets which may be flat or rounded. Chisholm and Dean (1974, p17) recognised upward fining cycles. The finer-grained muddy and silty beds are often greenish in colour when exposed.

Formerly included in the Upper Old Red Sandstone (Chisholm and Dean 1974) the Kinnesswood Formation has been more recently assigned to the Devono-Carboniferous Inverclyde Group by Paterson and Hall (1986), who comment on the two main lithologies present: (1) fine- to medium-grained, poorly bedded sandstone with greenish siltstones and mudstones and (2) coarse-grained, better cemented, sometimes pebbly sandstone with quartzite and vein quartz pebbles together with rip-up siltstone clasts and 'cornstone' clasts (see below). Erosive bases to the sandstones occur as do fining-upward sequences.

3. Kinnesswood Row: Kinnesswood Formation, 'cornstone'

Continue up the side of Kinnesswood Row as far as a 1.5m thick cornstone. This is light orange-brown in colour and is located 3m below and 10m south of the conspicuous 30cm thick coal seam. The cornstone is sandy and gnarled in appearance and is highly resistant to weathering – pieces of it can be picked up all the way to the bottom of the gully. It effervesces readily when treated with dilute hydrochloric acid. This bed is at the top of the Kinnesswood Formation and was taken by Geikie (1900, p35) to mark the top of the Old Red Sandstone. There is little doubt that it is a calcrete or caliche, a fossil soil profile typical of semi-arid areas in many parts of the world today. Paterson and Hall (1986) have used the occurrence of carbonate-bearing sediments, whether calcretes or cementstones, at or about this horizon to delineate their Inverclyde Group sediments with, here, the aeolian Knox Pulpit Formation beneath and the fluvio-deltaic Strathclyde Group above. Immediately above the cornstone is a 15cm thick coaly shale which is, however, rarely exposed. This is overlain by 3m of poorly exposed sandstone containing large roots or *Stigmaria* and the 30cm coal seam mentioned above. Browne (1980, p325) reported the recovery of an undoubted Carboniferous miospore flora from a 15cm coal near the base of the Strathclyde Group. On the evidence of the coal seam and plant fossils there is clearly a major environmental change in the conditions of sediment accumulation from a semi-arid fluviatile one to a humid fluvio-deltaic one. This is reflected too in the abrupt fall in feldspar content in the sandstones from 20 per cent below to 1–4 per cent above (Chisholm and Dean 1974, p4).

Nevertheless Paterson and Hall (1986, p8) tentatively place the Devonian-Carboniferous boundary at the base of the

Kinnesswood Formation rather than at what seems locally a more impressive change.

From this point onwards, the exposures, though generally sparse, are better on the northern side of the gully. They consist of a further 30m of Strathclyde Group sediments with dark grey shales predominating. The sandstones again form small ledges while the shales are generally grass covered. The succession is summarized in the table below:

SUCCESSION IN KINNESSWOOD ROW

		metres
	Quartz-dolerite sill	36.5
Lower	Sandstones with minor shales, seat-	
Limestone	earth and tuffaceous sandstone	12.0
Formation	Charlestown Station Limestone	3.0
	Mainly shales with crushed brachio-	
	pods and bivalves	6.5
	Coal seam	0.6
	Black shales with *Lingula*, seatearths,	
Undifferentiated	green shaly sandstones (fakes) and	
Strathclyde	one thin limestone band	15.0
Group	Sandstones	2.5
	Coal seam	0.3
	Sandstones with *Stigmaria*	3.0
	Thin coal seam (seldom exposed)	0.15
Inverclyde Group		
Kinnesswood	Cornstone	1.2
Formation	Sandstones and siltstones	60.0
Stratheden Group		
Knox Pulpit		
Formation	Aeolian sandstones (top only seen)	100.0

4. *Charlestown Station Limestone and base of quartz-dolerite sill*

The Charlestown Station Limestone, though it is no longer exposed in the quarry at the top of Kinnesswood Row, can be examined 25m north and about 6m lower on the hillside where its brown weathering and grey crystalline interior serve to distinguish it from the adjacent sandstones. The limestone crops out lower on the hillside because of a series of E–W faults which cut the outcrop and successively downthrow to the north for a total of about 9m.

Above the quarry at the head of Kinnesswood Row it will be seen that the exposed sequence, which extends up to the base of the Midland Valley Sill, consists predominantly of sandstone with minor shales, seatearths and some apparently tuffaceous sandstone. **CARE** should be taken when examining these outcrops because footing is not good and blocks of the sill are easily dislodged. The sediments have been thermally metamorphosed beneath the base of the sill, the topmost few metres of the sandstone having been altered to quartzite and the shale band bleached and hardened. The base of the quartz-dolerite sill is chilled against the underlying sediment, but no fresh specimens of the chilled margin are available. A few metres higher up, the sill shows crude columnar jointing and spheroidal weathering.

5. Old quarry: Charlestown Main Limestone, Lower Limestone Formation

Now continue north-eastwards obliquely up the hillside for a few metres to join a grassy track that ascends to the top of the scarp and follow the track for 180m to an old quarry in the Charlestown Main Limestone. Note that the sediments in this quarry lie above the sill. In the quarry 4m of dark grey shales, the basal 2m with common carbonate nodules, overlie the limestone of which only the top 50cm are exposed however. The shales are moderately fossiliferous yielding *Spirifer*, *Productus* and rarely goniatites. Looking south from the quarry the scarp formed by the E–W faults which cut the Charlestown Station Limestone on the face of Bishop Hill can be seen. The quartz-dolerite sill is faulted against the overlying sediments of the Lower Limestone Formation and stands up prominently on the southern or upthrow side of the fault.

FIGURE 18: Clatteringwell Quarry, Bishop Hill. The Charlestown Main Limestone, normally an alternation of limestone and shales as seen in the foreground, is increased in thickness by 5m on account of the massive algal reef seen in the cliff in the background.

6. Clatteringwell Quarry: Charlestown Main Limestone and reef

In Clatteringwell Quarry, 140m further east, the Charlestown Main Limestone is well exposed for 275m in old workings. The succession is as follows:

	metres
Thin-bedded limestone with local 5m thick reef development	7.0m
Limestone with thin shaly partings at the top	1.0+

Near the north-western end of the quarry face a reef of massive limestone up to 5.0m thick occurs within the top thinly bedded limestone. The shaly limestone has been stripped off over a short distance to reveal the form of the reef which appears to be about 50m wide and elongated 10° E of N. In thin section, however, little structure is visible in the reef rock; indeed the only fossils visible are crinoid ossicles and these have been recrystallised, and it seems probable that the reef builders were largely algal: certainly corals are rare. The rock is recrystallised lying as it does only a few metres above the major Midland Valley Sill. The succeeding shaly limestone is also recrystallised, fossils often weathering out. They include the brachiopods *'Spirifer'* and *'Productus'* together with *Aviculopecten* and *Fenestella*. Shales overlying this are exposed a few metres north-west of the reef. Fossils are most easily collected from the spoil heaps. At the south-eastern end of the quarry the dip of the strata is about 12° S, but no higher beds are exposed. Again the rocks are baked by the underlying sill – the limestone is recrystallised and the shale hardened so that the fossils now weather out.

7. Bishop Hill scarp face: view

One hundred metres north-west from Clatteringwell Quarry cross the wall at the gate and continue in the same direction to the scarp edge from which, weather permitting, there is a fine view to the north and west. The hills of North Fife and the Eastern Ochils from 6.5 to 16km away are composed of Lower Old Red Sandstone lavas and tuffs belonging to the Ochil Volcanic Formation and dipping south-east at 15° thus presenting their dip surfaces to the viewer. The lavas dip unconformably beneath the almost horizontal sediments of the Upper Devonian Stratheden Group which in turn pass up into the Devono-Carboniferous Inverclyde Group and the Strathclyde Group of the Carboniferous on the face of the Lomond Hills and Bishop Hill, there to be capped by the quartz-dolerite of the Midland Valley Sill. To the west Loch Leven occupies what is believed to be a very large kettle hole. It is surrounded by extensive fluvioglacial deposits and glacial till. To the south of the loch Benarty is also capped by the Midland Valley Sill and to the south-west the Cleish Hills display the same sill, much broken by faults, together with Lower Carboniferous sediments and volcanics.

During the later stages of the Devensian glaciation of the area east-flowing ice is thought to have divided round the Lomonds-Bishop Hill massif, one branch flowing down the

Firth of Forth to the south and the other down Stratheden to the north.

Now take the rough track which zig-zags down the face of Bishop Hill, noticing on the way the good view of Kinnesswood Row and the geological features seen on the way up the hill, especially the sill. The track ends in a field 350m behind Balnethill Farm. Return to the bus and then to St Andrews by retracing the outward route.

References

BROWNE, M. A. E., 1980. Stratigraphy of the Lower Calciferous Sandstone Measures in Fife. *Scot. J. Geol.*, **16**, 321–8.

CHISHOLM, J. I. and DEAN, J. M., 1974. The Upper Old Red Sandstone of Fife and Kinross: a fluviatile sequence with evidence of marine incursion. *Scot. J. Geol.*, **10**, 1–30.

GEIKIE, A., 1900. The geology of central and western Fife and Kinross. *Mem. Geol. Surv. Gt. Br.*

HALL, I. H. S. and CHISHOLM, J. I., 1987. Aeolian sediments in the late Devonian of the Scottish Midland Valley. *Scot. J. Geol.* **23**, 203–8.

PATERSON, I. B. and HALL, I. H. S., 1986. Lithostratigraphy of the late Devonian and early Carboniferous rocks in the Midland Valley of Scotland. *Rep. Br. Geol. Surv.* **18**, No 3.

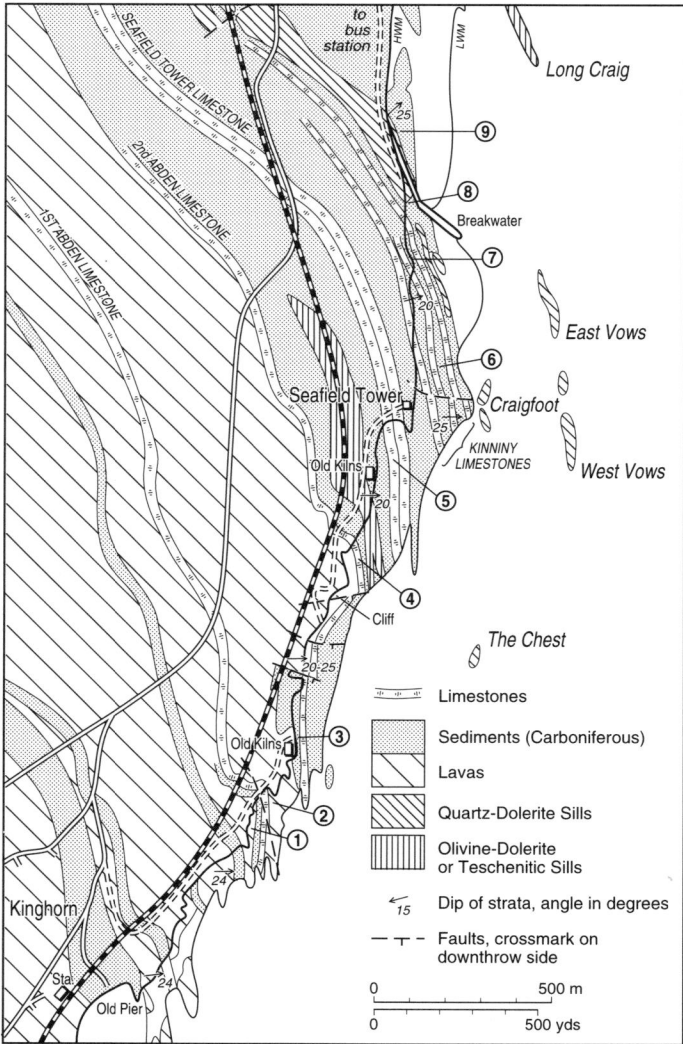

MAP 23: **Kinghorn - Kirkcaldy.**

272

Excursion 18

Kinghorn–Kirkcaldy (whole day)

OS 1:50,000, Sheet 66
GS One-inch Sheet 40
Excursion Map 23.

WALKING DISTANCE: 0.8km of paths and 2.5km on rocky shore.

PURPOSE: The main objectives of this excursion are to examine Carboniferous rocks belonging to (1) the Strathclyde Group, here predominantly basaltic lavas and tuffs with some inter-bedded sediments and of the order of 425m thick (Francis 1961); (2) the Lower Limestone Formation (highest Dinantian) 146m thick; and to a much lesser extent (3) the Limestone Coal Formation (Namurian). Intruded into these are (4) a teschenite sill which cuts the Lower Limestone Formation sediments and (5) a quartz-dolerite sill cutting the lowest sediments of the Limestone Coal Formation.

Throughout the section the strata dip east at 20°–30° off the Burntisland Anticline and into the Fife Coalfield syncline. The lavas of the Strathclyde Group are probably related to a series of vents exposed in the vicinity of Burntisland and Aberdour a few kilometres to the west. Age dates on these lavas are around 330My (Francis 1991, p396), contemporary with the last stages of the Clyde Plateau Lavas and belonging to the late Viséan.

ROUTE: The outward journey is by B939 through Ceres to Craigrothie, then by A916 to Kennoway and Windygates.

273

From here take the A915 to Kirkcaldy, along the promenade and by A921 to Kinghorn. In Kinghorn leave the bus at Ladyburn Place at the northern end of the town and send it back to the car park at the southern end of the Kirkcaldy promenade. The excursion is best undertaken on foot from Kinghorn to Kirkcaldy.

Walk down Ladyburn Place, under the railway to join the footpath running between the railway and the caravan site and follow this northwards down to the shore to see the top of the Strathclyde Group succession which comprises a series of olivine-basalt lava flows, mainly of 'Dalmeny' and 'Hillhouse' types (Macgregor 1928, Upton 1982, p273), with intercalations of sandstone, ashy sandstone and muddy sandstone (fakes).

1. Pillow structure

The highest lava in this part of the section displays what Geikie (1900, p72) called pillow structure. This is a product of weathering and jointing rather than true pillows formed during eruption under water. Indeed some of the flows show a degree of subaerial weathering. In the lava there is a concentric arrangement of vesicles, together with a crude columnar structure normal to the top and bottom of the flow. The main part of the flow comprises hard, blue-grey, compact basalt passing up into a softer, greenish, amygdaloidal basalt, many of the vesicles containing calcite. Notice also, on a flat weathered surface at the eastern end of the basalts, veins of calcite following the joints and forming a hexagonal pattern. This rock may be seen on the western side of a small bay, readily identified from the path when approaching from Kinghorn by means of the pale grey limestone on its eastern side.

2. First Abden Limestone, lavas and laterite

At this same locality the following succession can be made out:

	metres
Basalt lavas	—
Calcareous shale, contorted under the lavas	0.5
First Abden Limestone in several thick beds with shale partings between (pale grey at HWM)	3.5
Shelly grey shale with 2cm bentonite at base	0.15
Grey-green tuffs with graded bedding	1.6
Shelly grey shale with the Abden Bone Bed near the base	1.2
Seatearth with plant fragments	0.6
Basalt lavas with weathered top	—

Correlation of the First and the Second Abden Limestones with the succession in the Lower Limestone Formation elsewhere in Scotland has been controversial over the years. However, Wilson (1989, p104) has suggested that the Hurlet Limestone at the base of the Lower Limestone Formation is best correlated with the Second Abden Limestone and suggested further that the First and Second Abden Limestone may well be part of a single episode of limestone formation which was locally interrupted by eruption of the lavas which lie between them on the coast. Certainly the top of the thick lava of the Strathclyde Group lies immediately below the Second Abden Limestone in the shafts of the former Seafield Colliery some 2km to the north and there is no trace of the First Abden Limestone there (Francis 1961, p17).

A small N–S fault downthrowing to the west and repeating most of the succession should be examined where a small stream reaches HWM. Considerable fault drag of the strata can be seen. Next, the shales above and below the tuffs should be carefully examined as they are richly fossiliferous, yielding '*Poductus*', *Aviculopecten, Sanguinolites, Murchisonia,* ostracods, occasional '*Orthoceras*' and many other fossils. The shales above the graded tuffs have at their base a 2cm yellow bentonite band. The shales interbedded in the limestones are also very fossiliferous as is the limestone itself, but in the latter the fossils are very difficult to extract and are better examined on the wave-washed surfaces. Foraminifera are common in thin sections of the limestone. Notice that the shale above the limestone is squeezed and contorted where the overlying basalt came into contact with it while it was still soft and water saturated. This shale has been selectively eroded out by the sea and forms a fissure at the foot of the lava cliff.

Now return to the footpath and, keeping close to the shore, continue north, noting that the basalt lava flow above the limestone is quite massive. On top of this flow is a 1.2m thick green tuff band now largely altered to a brick red laterite in which small pieces of decomposed lava can still be found. This laterite lies at the base of a prominent scarp formed by the next lava flow which is massive, doleritic and displays carious weathering along the joints. In all there are some 21m of strata, mainly lavas, separating the two Abden Limestones (Francis 1961, p23).

3. *Old kilns: Second Abden Limestone; faulting on the Abden shore*

Beyond the two old kilns, formerly used for burning lime quarried from the Second Abden Limestone which crops out on the shore here, the path ascends to beside the railway. Leave the path and follow the old path down to the shore where the following succession is exposed

	metres
Sandstone and shaly sandstone	—
Second Abden Limestone	4.0
Fossiliferous, dark-grey, calcareous shale	2.75
Naiadites crassa band	0.02
Seatearth	0.6
Tuffs passing up into red laterite	6.0
Vesicular top to basalt lavas with pillow structure	—

The pillow structure in the underlying basalt is best exposed at HWM. Green tuffs appear to have penetrated the cracks in the upper surface of the lava before covering it completely. On the shore the tuffs, red laterite and seatearth, have been preferentially eroded by the sea, but in the low cliff at HWM they form conspicuous coloured bands. The succeeding grey shales were described in detail by Ferguson (1962) who recognised from bottom to top the stages of a marine transgression, documented by the changing very abundant fauna. At the base will be seen the hard, 2.5cm *Naiadites crassa* band in which many of these thick-shelled bivalves are found. The next 90cm of shales yield in particular *Lingula* and *Streblopteria*. The next 75cm yield *Schizophoria* and *'Productus'* and the topmost 60cm, immediately beneath the limestone, yield *'Productus'*, corals and bryozoa. On the shore the Second Abden Limestone forms a prominent ridge which decreases in height when followed north. It consists of a series of limestone beds separated by irregular muddy partings. Considerable colonies of *Lithostrotion junceum* occur in it and solitary corals can also be seen, particularly on the wave-washed upper surface. Crinoid debris is abundant throughout.

For the next 550m the Second Abden Limestone strikes parallel to the shore but is cut by three small dip faults which displace the outcrop dextrally, in each case by a small downthrow to the south. At the first of these faults, 150m north of the kilns, the limestone is brecciated, dolomitised, brown in colour and has a hematitic stain.

At the second fault, 150m further north, there is evidence of slight tear movement and again breccia is present. The third fault, lying just south of the prominent cliff of lava, cuts the limestone without appreciable brecciation.

4. Second Abden Limestone; teschenite sill

The footpath now skirts the western end of the cliff beside the railway line, then runs along the top of the cliff before descending to the shore beyond the headland formed by the lava cliff. (At low tide it is possible to pass the seaward end of the cliff.) The Second Abden Limestone, together with the beds beneath, should be carefully examined here and compared with the section seen at Locality 3; it will at once be noticed that the succession, which is tabulated below, is much thinner, principally on account of the tuffs beneath the limestone being much thinner.

	metres
Sandstone and sandy shales	—
Second Abden Limestone	5.0
Fossiliferous dark grey shales	1.2
Naiadites crassa band	0.02
Seatearth, passing down into tuffs	0.15
Green tuffs	1.2
Basalt lavas	—

The metal footbridge on the path traverses an E–W fault line with slight brecciation in the lavas along which a cave has been opened up by the sea. The ridge on the shore formed by the Second Abden Limestone has also been breached, although the fault displacement is only about 1m. Northward it will be seen that the Second Abden Limestone is overlain by 25m of sandstones and shaly sandstones, within the outcrop of which the shore is crossed by a teschenite sill forming a small scarp on the shore. The sill margins are bleached and altered to white trap and the adjacent country rock has been baked. At the base of the sill at one location a set of tension gashes around 30cm long and filled with calcite are exposed. The transgressive nature of the sill and its southward splitting are displayed on the shore. Notice the overlying white and pink sandstones which show good cross bedding and ripple marking. Interbedded seatearths contain abundant *Stigmaria*. They are succeeded by 3m of grey shales which become shelly upwards and pass up into the Seafield Tower Limestone.

5. *Old Kilns: Seafield Tower Limestone*

The Seafield Tower Limestone, at one time worked on the shore, crops out as a prominent pale grey band on the shore to the east of four old lime kilns, 250m south of Seafield Tower. The limestone, which has three main beds, is 3.5m thick and is succeeded by 14m of calcareous grey shales with carbonate nodules and bands of crinoidal limestone up to 1m thick. All

FIGURE 19: Seafield Tower, Kirkcaldy. This stands on a 15m thick sandstone. Beneath this lies a sequence of shales and limestones, in particular the Seafield Tower Limestone which forms the conspicuous pale ridge in the foreground.

these calcareous strata have been grouped together on the map as the Seafield Tower Limestone. *Lithostrotion, 'Zaphrentis', 'Productus', 'Spirifer', 'Athyris', Schizophoria,* crinoids and many other fossils can be found in these beds in considerable numbers.

Forsyth (1970, p8) correlated the Seafield Tower Limestone with the Charlestown Main Limestone elsewhere in Fife, although Wilson (1966, p113) noted that the unique Neilson Shell Bed fauna, widespread elsewhere above the Charlestown Main Limestone and its equivalents, had not been located above any of the limestones of this section.

6. Craigfoot: Seafield Tower, Kinniny Limestones

Seafield Tower stands on the outcrop of a prominent cross-bedded sandstone, 15m thick and locally reddish in colour. On its seaward side this sandstone is succeeded by 55m of sandstones, shales and thin limestones, the last of these aggregating some 5m and forming small scarps on the shore. These limestones which are the equivalent of the Hosie Limestones elsewhere in Scotland (Wilson 1989, p95) are the Kinniny Limestones, the highest of which marks the top of the Lower Limestone Formation in Fife. The succession is listed below in generalised form (after Francis 1961, p18):

	metres
Upper Kinniny Limestone	0.5
Shales and thin sandstones, shelly at base; sandstone increasing to north	6.5
Limestone	0.15
Shales	0.6
Limestone, sandy	0.5–1.4
Shales, shelly at top and bottom, with calcareous nodules	12.0
Middle Kinniny Limestone	2.3
Shales, coaly at top and with subsidiary sandstones	7.0
Sandstone	3.5
Shales, sandy towards the base	4.75
Lower Kinniny Limestone	0.7
Shales with sandstones, shaly sandstones, a 90cm coal seam and two thin limestone bands	15.0
Thick trough cross-bedded sandstone, on which stands Seafield Tower	15.0

In this succession the coaly beds are rarely exposed though the coal beneath the Lower Kinniny Limestone was at one time worked at outcrop. The Lower Kinniny Limestone is bioturbated, sandy and contains nodules up to 45cm long by 15cm thick. The Middle Kinniny Limestone is conspicuous on account of its being cut by several small thrusts which die out quickly in the shales above and below. It is richly fossiliferous in its upper, muddier part. The 1.4m limestone above the Middle Kinniny Limestone also forms a prominent feature on the shore, especially towards Craigfoot, while the 12m thick shale between these two limestones has been eroded out leaving an intervening depression. The 0.5m thick Upper Kinniny Limestone is impure, passes down into sandstone and thins

FIGURE 20: Minor thrusting in the Middle Kinniny Limestone, Seafield shore, Kirkcaldy. Offshore the dark rocks, the East Vows, are part of a dolerite sill.

northwards. It is stratigraphically important since it forms the topmost bed of the Lower Limestone Formation. Offshore the crags of Craigfoot and the East and West Vows are formed by dolerite sills intruded into the overlying Limestone Coal Formation sediments.

7. *Quartz-dolerite sill: sedimentary structures*

From Craigfoot follow the Upper Kinniny Limestone north along the shore for 400m until the outcrop of the sill which forms Craigfoot is reached. The sill, which is 5m above the base of the Limestone Coal Formation, i.e. the top of the Upper Kinniny Limestone, here forms a scarp and is about 3m thick. The margins are bleached to white trap and are fine grained. Along this part of the shore it will be seen that the strike gradually swings round from N–S to NNW–SSE but the dip remains at 20°–25° eastward. Two hundred and thirty metres south of the bend in the old Tyrie breakwater some of the sandy beds beneath the Upper Kinniny Limestone have developed pseudo-nodules, through pods of sandstone sinking into the siltstone beneath while the sediments were still un-consolidated. Other thin-bedded sandstones in the vicinity display 'Scolicia' type tracks across the bedding planes, gen-erally thought to be made by gastropods. A small E–W fault with a downthrow of around 3m can also be seen at this locality.

8. *Tyrie breakwater: white trap sill*

Sixty-five metres south of the bend in the breakwater the quartz-dolerite sill that forms Craigfoot disappears under the sand. It reappears 27m further north where a large brick pipe crosses the shore. Here too the sill margins can be seen to be fine grained and altered to white trap. A few metres north of the pipe a large raft of sandstone, now baked to quartzite and with its bedding perpendicular to the sill margins, is plainly visible in the sill.

9. *Limestone Coal Formation*

Now walk to the northern end of the breakwater where the coast section can be examined once more. The outcrops here are mainly of cross-bedded sandstones together with muddy sandstones and thin beds of shale with ironstone nodules, belonging to the Limestone Coal Formation. The quartz-dolerite seen at the bend in the breakwater also crops out on its seaward side and is again altered to white trap. Continue along the shore to the car park to rejoin the bus and return to St Andrews.

References

FERGUSON, L., 1962. The paleoecology of a Lower Carboniferous marine transgression. *J. Paleont*, **36**, 1090–1107.

FORSYTH, I. H., 1970. Geological Survey boreholes in the Lower Carboniferous of West Fife. *Bull. Geol. Surv. G.B.*, **31**, 1–18.

FRANCIS, E. H., 1961. Economic Geology of the Fife Coalfields, Area II, Cowdenbeath and Central Fife. *Mem. Geol. Surv. Scotland.*

————, 1991. Carboniferous-Permian igneous rocks. In

G. Y. Craig (Ed.) *Geology of Scotland,* pp393–420. Geological Society, London.

GEIKIE, A., 1900. The geology cf Central and Western Fife and Kinross. *Mem. Geol. Surv. Gt. Br.*

MACGREGOR, A. G., 1928. The Classification of the Scottish Carboniferous olivine-basalts and mugearites. *Trans. Geol. Soc. Glasg.* **18**, 324–50.

UPTON, B. J. G., 1982 Carboniferous to Permian volcanism in the stable foreland. In D. S. Sutherland, (ed.) *Igneous rocks of the British Isles,* pp 255–75 Wiley-Interscience.

WILSON R. B., 1966. A Study of the Neilson Shell Bed, a Scottish Lower Carboniferous marine shale. *Bull. Geol. Surv. Gt. Br.,* **24**, 105–30.

————, 1989. A study of the Dinantian marine macrofossils of central Scotland. *Trans. Roy. Soc. Edinb., Earth Sci.,* **80**, 91–126.

Index